RAND PROJECT AIR FORCE

T0303213

Estimating Air Force Deployment Requirements for Lean Force Packages

A Methodology and Decision Support Tool Prototype

Patrick Mills, James A. Leftwich, Kristin Van Abel, Jason Mastbaum

Prepared for the United States Air Force
Approved for public release; distribution unlimited

For more information on this publication, visit www.rand.org/t/RR1855

Library of Congress Control Number: 2017959648
ISBN: 978-0-8330-9754-5

Published by the RAND Corporation, Santa Monica, Calif.

Preface

The Air Force has been grappling for several years with how to survive and fight in contested, degraded, and operationally limited (CDO) environments, and one of its recent innovations has been the advancement of basing concepts that require significant resilience and mobility of combat forces. These concepts are still under development, and the need for mobility and agility places pressure on planners to reduce the military footprint and potentially take significant risks in the interest of speed.

The Air Force does not currently have a comprehensive tool or methodology for integrated deployment planning that can rapidly explore trade-offs among capability (or risk), speed, and cost to achieve lean force packages for use in CDO environments. The purpose of this analysis is to describe and demonstrate a methodology and prototype tool for lean force package planning and analysis—called the Lean Strategic Tool for the Analysis of Required Transportation (Lean-START)—that does just that.

Lean-START, an Excel-based spreadsheet model, determines the list of equipment and personnel required to support a user-specified operation, along with the movement characteristics of the materiel for a wide range of support areas. It acts as both a demand generator of the manpower and materiel needed at an expeditionary base to achieve initial operating capability, and also an iterative planning environment to inform course of action and concept development.

The research reported here was commissioned by the U.S. Air Force and conducted within the Resource Management Program of RAND Project AIR FORCE (PAF) as part of a fiscal year 2016 project entitled "Enhancing Global Agile Combat Support (ACS) Processes." The work was sponsored by the Commanders of the Air Force Installation and Mission Support Center and the Air Force Sustainment Center and should be of interest to personnel involved in logistics, installation and mission support, or sustainment in the U.S. Air Force, in particular logistics planners and those on Air Force forces staffs.

To receive a prototype version of the tool, please contact Patrick Mills at pmills@rand.org, or contact the Director of PAF's Resource Management Program by visiting www.rand.org/paf/contact.

RAND Project AIR FORCE

RAND Project AIR FORCE (PAF), a division of the RAND Corporation, is the U.S. Air Force's federally funded research and development center for studies and analyses. PAF provides the Air Force with independent analyses of policy alternatives affecting the development, employment, combat readiness, and support of current and future air, space, and cyber forces. Research is conducted in four programs: Force Modernization and Employment;

Manpower, Personnel, and Training; Resource Management; and Strategy and Doctrine. The research reported here was prepared under contract FA7014-06-C-0001.

Additional information about PAF is available on our website: www.rand.org/paf/

Contents

Figures

Tables

Summary

In the early 1990s, the U.S. Air Force embarked on an effort to be more responsive to worldwide demands for airpower. The intent was to create an Expeditionary Aerospace Force (EAF) that could provide a lean, agile airpower capability with short response times across the world.[1] The EAF capabilities would allow the Air Force to stabilize the operations tempo of the force by using rapid-response capability in lieu of a permanent forward presence for deterrence.[2] In response to the challenge, the Air Force developed lean force-deployment packages that—when complemented by a preestablished, robust, worldwide infrastructure of war reserve materiel and forward-positioned aircraft-repair facilities—could provide desired airpower in the time needed.

Today, motivated by a different set of issues, the Air Force finds itself needing to create even leaner force packages capable of providing a full spectrum of operational effects while being mobile. The United States now faces threats from near-peer adversaries. Those adversaries are capable of creating contested, degraded, and operationally limited (CDO) environments within which the Air Force must operate.[3] The emerging threats and security environments have spawned new operational concepts that require the Air Force to be both more resilient and more mobile in order to survive an attack and deliver airpower after an attack. The speed and agility originally demanded as characteristics of the EAF for initial response are now required of forces already deployed and operating in a CDO environment.

The challenge of planning for these operations is compounded by the fact that the Air Force rarely deploys anything resembling a standard force package, requiring planners to tailor each package to suit a commander's intent and existing base capabilities. To address these challenges, PAF developed a tool prototype for planning and analysis, called the Lean Strategic Tool for the Analysis of Required Transportation (Lean-START).

---

[1] The term *Expeditionary Aerospace Force* was used to describe the emerging operational concept that focused on the Air Force being a force capable of rapidly deploying anywhere in the world. The concept was implemented in the Air Force by organizing the force into Aerospace Expeditionary Forces (AEFs), Aerospace Expeditionary Wings, Aerospace Expeditionary Groups, and Aerospace Expeditionary Squadrons. The primary deployment package was a collection of unit capabilities called the AEF. Air Force units were parsed into AEFs and placed on a rotational deployment schedule that also added stability and certainty to the deployment schedules for units. We focus on the employment concept embodied in the EAF.

[2] Robert S. Tripp, Lionel A. Galway, Paul Killingsworth, Eric Peltz, Timothy Ramey, and John G. Drew, *Supporting Expeditionary Aerospace Forces: An Integrated Strategic Agile Combat Support Planning Framework*, Santa Monica, Calif.: RAND Corporation, MR-1056-AF, 1999, p. 3.

[3] These operating environments are also sometimes referred to as anti-access area denial (A2AD).

The purpose of this report is to describe the basic philosophy and operation of Lean-START, use it to demonstrate basic calculations of lean force packages, and then to illuminate some of the basic dynamics and drivers of lean deployments.

To receive a prototype version of the tool, please contact Patrick Mills at pmills@rand.org, or contact the Director of PAF's Resource Management Program by visiting www.rand.org/paf/contact.

Planning and Risk

The U.S. Department of Defense (DoD) and the Air Force have adjusted their planning processes to deal with the dynamics of the emerging security environment. The changes are embodied in a planning framework called Adaptive Planning and Execution (APEX). Two of the fundamental characteristics of the APEX framework are (1) the focus on iterative development and analysis of different courses of action (COAs) to achieve the desired objectives of the campaign and (2) the continuous assessment of risks in each COA.

The DoD planning process is generally practiced by a combatant commander's (CCDR's) planning staff and includes opportunities for input by the component planning staffs. As CCDR planners are considering and analyzing various COAs, they take inputs from the Air Force component planning staff, which typically conducts planning concurrently but independent of the CCDR's team. Both planning teams conduct planning and assess risks iteratively.

The dialogue between the two teams must be informed by analysis and data that consider the trade-offs that are made among capability, speed, and risks. Lean-START is designed to help Air Force planners provide the data and support the discussions with the CCDR's planning team. Lean-START can quickly demonstrate how to increase speed of deployment by adjusting the designed sortie production capability of the deploying force. Likewise, it demonstrates the trade-off between speed of deployment and risk-mitigating capabilities (e.g., firefighting).

Joint policy recognizes that addressing risk in military operations is more of an art than a science. There are difficulties in determining the probability that a potential risk will become reality. There are difficulties in estimating the negative impact of a potential risk. While Lean-START cannot definitively address the probability and impact of a risk, it does provide a framework for considering trade-offs between risk tolerance and speed of deployments, dictated by the size of the force *footprint* (the sum of the personnel and amount of equipment).

To increase the speed of deployment, planners can reduce the size of the deploying package by eliminating equipment or personnel. The planners might preposition the eliminated items closer to the point of use, thereby reducing the time required to move them; acquire the items at the deployed location by contracting for them on the local market or making arrangements with the host nation; or decide to operate without the items. By eliminating resources from a deployment package and not replacing them, planners are creating known shortfalls equating to some degree of capability and introducing a risk to operations. For the purpose of Lean-

START's risk analysis, we categorize resources, and accordingly risks, into four categories: mission capacity and duration, safety, security, and human performance and endurance. These classifications allow planners to consider risk tolerance of reducing a capability by considering the part of the operation that will face negative effects if the risk comes to fruition.

Design of the Tool

Lean-START is an Excel-based spreadsheet tool that operates in essentially four stages. The first stage is the input for the operational scenario: duration, type and number of aircraft, and more-detailed mission information. These inputs drive demands for several areas:

- **aviation:** pilots and other mission duties
- **maintenance:** flight-line and repair personnel working directly with the aircraft, its components, and aerospace ground equipment
- **munitions maintenance:** personnel to account for, build up, and transport munitions (chaff, flare, missiles, and bombs) to the flight line
- **munitions:** the actual weapons themselves.

The second stage is beddown planning, which provides a range of inputs for nonmaintenance agile combat support (ACS). These inputs provide the user with options for both the inherent characteristics of the destination location and the desired level of capability for the operation (for example, fuels, security forces, and firefighting).

The third phase is tailoring. There are multiple screens with which the user can tailor the requirements, based on information or assumptions about existing levels of support provided by prepositioned equipment, host-nation support, contract support, or risk tolerance.

The fourth phase is outputs. These include detailed lists of personnel, equipment, and unit type codes on a number of screens, as well as aggregate movement characteristics in the form of personnel, equipment, and transportation requirements. Finally, given those outputs, the user can iterate to change any of the above inputs to explore the tradespace of options.

Illustrative Applications

In the report, we demonstrate some of the model's capabilities with several illustrative calculations. We highlight some of the key drivers of footprint, discuss trade-offs with distributed basing, demonstrate time-phasing of deployment requirements, and show the use of war reserve materiel (WRM) prepositioning to offset footprint. We also compare the drivers of footprint to their levels of risk.

Lean-START supports the need for speed and agility of both force employment and planning in a number of ways. First, it allows planners to quickly generate requirements for a scenario. Second, it provides analysis to illuminate the trade-offs among speed, capability, and risks. Finally, it informs COA development and selection.

Using Lean-START, planners are able to quickly determine the combat support capabilities required for a specified operational scenario. By identifying such operational parameters as the number and types of aircraft, sortie rates for those aircraft, duration of operations, and types of missions being flown, Lean-START will generate information about the quantities of support resources required, using traditional planning factors. This is a capability that existed in Lean-START's predecessor, Strategic Tool for the Analysis of Required Transportation (START).

Lean-START also provides planners with an opportunity to examine trade-offs. How can speed of deployment be increased by modifying the operational force package or required sortie production capability of the deploying weapon systems? How can speed of deployment be increased by accepting risk to the sortie production capability or to the safety and security of the deploying units? What resources that support care and feeding of deployed forces can be eliminated to improve speed? These are questions that get at the fundamental questions of trading capability and risks for speed of deployment.

Finally, as the Air Force conducts planning in support of a CCDR's campaign, Lean-START enables rapid iteration of COAs that can be provided to the CCDR's planning team. As highlighted in Joint Publication 5-0,[4] planners often must consider and analyze several COAs concurrently. This dynamic planning and concurrent replanning is a fundamental element of the adaptive planning and execution framework, and it is supported by Lean-START's ability to quickly compute deployment requirements and address trade-offs.

Limitations of the Model

We find two key limitations of the model. For aircraft maintenance, it is difficult to codify in discrete terms many decisions that require complicated calculations about system interactions. Sizing and shaping a maintenance operation with personnel and equipment in the real world requires not two or three inputs, but perhaps a dozen or more—which begins to defeat the express purpose of simplicity in the tool.

The second limitation is in areas that require human judgment about risk, such as security forces. Assessing risk in a situation depends, again, on many factors, and after a certain point, only informed human judgment should be used to make the call.

In both cases, however, Lean-START succeeds in bounding the problem. It imagines stark breakpoints where continuous variables really exist, but it can serve to narrow the terms of a debate and explore input parameters in a way that help inform and calibrate human judgment, where the Lean-START user lacks deep subject matter expertise in one area or another.

---

[4] Joint Publication 5–0, *Joint Operational Planning,* U.S. Department of Defense, August 11, 2011, pp. IX–X.

Observations from Initial Lean-START Analysis

The analysis presented in this report is intended to demonstrate the range of calculations that the model can perform. Nonetheless, we do present some observations from our analysis.

One important question about lean force packages is where the breakpoints occur. For maintenance, the key transition is between one and two sorties and more-sustained sortie generation that requires more-robust repair capabilities. Other parameter changes bring increases in requirements (e.g., sortie rate and duration, extended deployments), but the biggest jump seems to be the shift to real repair operations, which require additional equipment, more personnel for multiple shifts and broader knowledge, and far more-robust spare-parts kits.

While nonmaintenance ACS is driven by a range of factors, duration drives a significant portion of the footprint, primarily due to basic expeditionary airfield resources. When operations are only focused on combat operations and/or limited to duration of a week or less, the force footprint drops drastically. Most of the equipment burdens for short- or long-duration operations are good candidates for WRM prepositioning, but that obviously comes with a cost.

The biggest single potential safety and security driver of footprint is security forces—both the personnel and equipment, particularly vehicles. With shorter durations, commanders may be more comfortable accepting more risk. But there are also myriad ways to address the need for security, including host-nation support; such new concepts as "fight the base;" and new technologies, such as unmanned aerial vehicles.[5]

Our analysis also has implications for operating in CDO environments more broadly. One of the outcomes for emerging basing and operating concepts is what we have already described: a kind of envelope within which requirements stay fairly small and outside of which requirements tend to balloon. This understanding can shape approaches to evolving support concepts.

Perhaps more significant is the need to view bases as part of an integrated network. Mobility air forces already do that, using the global en-route structure. It is business as usual for cargo and tanker aircraft to transit from location to location across a well-understood global network of bases. But given the potential size of footprint for combat forces—incorporating significant maintenance and munitions maintenance requirements, even within fairly short durations—much investment of time and perhaps money will be necessary to prepare and manage such a network of bases, even just a regional one.

Current operating concepts are now pushing combat forces to move in this direction. But that cannot be done without more-detailed understanding of base capabilities, needs, and spin-up time. It is also not possible without excellent coordination of host-nation support, contract support, and equipment prepositioning, with all elements integrated into a coherent strategy. The Air Force is heading in this direction, but much remains to be done. A tool like Lean-START can help.

---

[5] Bradley D. Spacy, Edwin H. Oshiba, and Nicholas J. Thomas, "Fight the Base, Recover the Base, Win the War!" *Air and Space Power Journal*, Vol. 30, No. 3, Fall 2016.

Recommendations and the Way Forward

Lean-START provides a relatively complete capability for determining deployment requirements and assessing various options for trading capability and risks for speed. However, additional areas could also be addressed that would benefit the Air Force.

- **Timing:** One variable that Lean-START does not currently incorporate is timing. The process of preparing, packing, deploying, and initiating operating capability is itself a highly coordinated and orchestrated operation, one with hard constraints and some insurmountable delays. In our discussions with subject-matter experts, timing was a common theme. This needs attention to better flesh out issues associated with packing up and leaving a base, as well as those associated with setting up operations at a base that starts cold—or even warm, with some presence and prepositioned equipment.
- **Operational contract support (OCS):** The Air Force has made significant improvements in the way OCS is factored into planning. In its current form, Lean-START can help target OCS operations by highlighting the larger items in a deployment package that could be tailored from the package and provided by OCS. Similarly, future iterations of Lean-START could expedite the planning process by providing the option to identify those items known to be available through OCS as part of the scenario creation.
- **Joint basing:** Often, the Air Force is identified as the responsible agent for Base Operating Support Integration (BOSI). In its current form, Lean-START only addresses installation support for Air Force organizations that are bedded down at a location. Future iterations could consider the requirements for joint or coalition forces that the Air Force is required to support at the deployed location.
- **Tactics, techniques, and procedures (TTPs) and policy:** Lean-START calculations are driven by current TTPs and planning factors. Lean-START also provides an opportunity to examine the effect of those planning factors on the total footprint for force support. The Air Force might consider evaluating current planning factors to determine if they should be adjusted to establish a new baseline for standard levels of support for combat operations.
- **Research and technology:** Lean-START provides the Air Force insight regarding which equipment dramatically increases the deployment footprint and thereby slows speed of deployment. Those items provide a target of opportunity for research and technology development. The Air Force might consider identifying some of those items and establishing a portfolio of research initiatives designed to provide the same capability at a lower cost to the deployment footprint.

1. Introduction

In the early 1990s, the U.S. Air Force embarked on an effort to be more responsive to worldwide demands for airpower. The intent was to create an Expeditionary Aerospace Force (EAF) that could provide a lean, agile airpower capability with short response times across the world.[6] The EAF capabilities would allow the Air Force to stabilize the operations tempo of the force by using rapid-response capability in lieu of a permanent forward presence for deterrence.[7] In response to the challenge, the Air Force developed lean force-deployment packages that—when complemented by a preestablished, robust, worldwide infrastructure of war reserve materiel and forward-positioned aircraft-repair facilities—could provide desired airpower in the time needed.

During the formative stages of the EAF, RAND Project AIR FORCE (PAF) provided analyses of the implications of the EAF concept on agile combat support (ACS). The following from an early report highlighted the effect of the EAF on ACS.[8]

> Because of the demand to respond quickly to contingencies by deploying from continental United States (CONUS), the new concept presents significant support challenges. The requirement to deploy quickly places pressures on the deploying units to minimize the amount of support they deploy; the demanding employment scenarios place counterbalancing pressures on the support system to ensure that there are sufficient resources to sustain combat operations.

Additionally, PAF's early analyses proposed several investments in infrastructure and equipment that would make an EAF concept feasible.[9]

> To support expeditionary operations, the Air Force will require a global infrastructure system consisting of Forward Operating Locations (FOLs) from which missions are flown, and Forward Support Locations/CONUS Support Locations (FSLs/CSLs)—regional facilities providing a selected set of support resources—and both location types must be assured resupply and a logistics command and control (C2) system to coordinate and tie together the FOLs and

[6] The term *Expeditionary Aerospace Force* was used to describe the emerging operational concept that focused on the Air Force being a force capable of rapidly deploying anywhere in the world. The concept was implemented in the Air Force by organizing the force into Aerospace Expeditionary Forces (AEFs), Aerospace Expeditionary Wings, Aerospace Expeditionary Groups, and Aerospace Expeditionary Squadrons. The primary deployment package was a collection of unit capabilities called the AEF. Air Force units were parsed into AEFs and placed on a rotational deployment schedule that also added stability and certainty to the deployment schedules for units. We focus on the employment concept embodied in the EAF.

[7] Robert S. Tripp, Lionel A. Galway, Paul Killingsworth, Eric Peltz, Timothy Ramey, and John G. Drew, *Supporting Expeditionary Aerospace Forces: An Integrated Strategic Agile Combat Support Planning Framework*, Santa Monica, Calif.: RAND Corporation, MR-1056-AF, 1999, p. 3.

[8] Tripp et al., 1999, p. xiii.

[9] Tripp et al., 1999, p. xiv.

> FSLs. This is a quite different system from either the system used during the Cold War or the austere "bring-it-all-from-CONUS" system implicitly envisioned during early discussions of the EAF. This new ACS system will require new planning methods as well.

The transition to the EAF was followed by a focus on capability-based planning (as opposed to threat-based planning) in the late 1990s and early 2000s. As with the EAF, there was a need to compose force packages and their logistics support quickly and deploy them on short notice.

Today, motivated by a different set of issues, the Air Force finds itself needing to create even leaner force packages capable of providing a full spectrum of operational effects while being mobile. The United States now faces threats from adversaries capable of creating contested, degraded, and operationally limited (CDO) environments, within which the Air Force must operate.[10] The emerging threats and security environments have spawned new operational concepts that require the Air Force to be both more resilient and more mobile in order to survive an attack and deliver airpower after the attack. The speed and agility originally demanded as characteristics of the EAF for initial response are now required of forces already deployed and operating in a CDO environment.

In those cases, the standard force packages are entirely too large to be practical (especially when rapid movements are called for). In response to the emerging operational concepts dictated by an enemy operating in a CDO environment, the Air Force asked RAND PAF to assess the implications of new operational concepts on the installation and mission support (I&MS) enterprise, including the following questions:

- How do the agile, deployable, dispersible employment concepts affect the demand for I&MS resources?
- What is the capability of the I&MS enterprise to execute a dispersed operations concept?
- What are the costs, benefit, speed, and risk trade-offs in executing dispersed operations?

PAF developed an analysis and decision support tool prototype called the Lean-Strategic Tool for the Analysis of Required Transportation (Lean-START) that endeavors to help answer those questions. As part of a larger PAF body of research, the tool aims to provide Air Force leadership with recommendations on how best to enable concepts for operating in a CDO environment.

To receive a prototype version of the tool, please contact Patrick Mills at pmills@rand.org, or contact the Director of PAF's Resource Management Program by visiting www.rand.org/paf/contact.

Although the security environment has changed over the years, there have been some enduring characteristics. From the early formulation of the EAF to today, the Air Force has remained an expeditionary force; it deploys to forward bases, performs operations, and redeploys

---

[10] These operating environments are sometimes referred to as anti-access area denial (A2AD).

to home stations. The forward bases have widely varying degrees of preexisting infrastructure and support, requiring planners to create highly tailored support packages. Operational forces are also highly tailored, only deploying the types and numbers of aircraft required by combatant commanders (CCDRs). These factors have driven the Air Force to develop a highly modular set of force packages, called Unit Type Codes (UTCs).

A UTC is a unit of capability specified by required manpower and equipment. UTCs range considerably in size: Some consist of a single individual with specified skills (e.g., a chaplain); others include dozens of personnel or hundreds of tons of equipment. UTCs are sufficiently small, modular, and numerous that sets of them can be assembled to express virtually any desired deployment capability that the Air Force requires.[11]

In recent years, the U.S. military has been called upon to perform more than 80 operations in dozens of countries worldwide, many of which fell outside the scope of its deliberate plans, and some of which it was called upon to do with minimal planning time. But over the course of these many operations, no standard Air Force deployment emerged. Each time a new deployment is anticipated, planners tailor force packages to fulfill their mission effectively but efficiently. In many cases, UTCs must be decomposed into their component personnel and equipment to right-size the deployment *footprint*, or the sum of the personnel and amount of equipment. Many UTCs are being pushed up to and beyond their limits for tailoring.

Flexibility is a strength of the Air Force, enabling it to provide just the right types and amounts of combat power and support in a timely manner. But paring and tailoring force packages (and their supporting personnel and equipment) to minimize footprint—particularly to preserve the scarce resource of airlift that enables such responsiveness—levies a significant burden on both planning and analysis processes. Planners must iteratively determine a support footprint for competing courses of action (COAs) to provide commanders with realistic options, balancing footprint and speed of deployment with mission assurance and risk. At the same time, analysts must explore numerous what-if situations to answer questions about operational scenarios and must explore trade-offs among different COAs to guide investments and policy.

The Air Force does not have a standard tool or methodology to explore various operational scenarios or evaluate COAs. In most cases, teams of subject-matter experts (SMEs) are assembled to develop deployment packages, a process that can absorb significant time and manpower and only analyzes one COA at a time. (We discuss this planning process more in later chapters.)

---

[11] The UTC is a U.S. Department of Defense (DoD) concept, mostly used in the context of deployment transportation data. The Air Force, however, has taken the UTC to its extreme in terms of granularity and modularity. DoD documentation defines a *UTC* (or type unit) as "a type of organizational or functional entity established within the Armed Forces and uniquely identified by a five-character, alphanumeric code." Joint Publication 1–02, *Department of Defense Dictionary of Military and Associated Terms*, U.S. Department of Defense, November 8, 2010.

Years ago, as part of the original body of analysis in support of the emerging EAF concept, PAF developed a tool to address this challenge, called the Strategic Tool for the Analysis of Required Transportation (START).[12] It employs a parameterized, rules-based algorithm to generate a list of UTCs that are necessary to support a user-specified deployment capability. Needed resources are specified as UTCs, depending on the characteristics of a base, the threat to which the base is exposed, and the numbers and types of aircraft at the base.

START uses traditional Air Force planning factors, which tend to default to robust, sustained operations. START was always intended as a starting point for planning and analysis, with later tailoring to be done based on situation-specific details. START has been used to support analyses on a number of topics, including AEF rotations,[13] war reserve materiel (WRM) prepositioning,[14] and, more recently, combat operations in a denied environment (CODE).[15]

Because of this renewed need for detailed tailoring in support of emerging basing concepts, PAF developed Lean-START. This tool for planning and analysis operates similarly to the original START model, but adds input parameters to help a user further tailor deployment packages, in many cases turning former assumptions (e.g., duration of the operation) into user-input variables.

Overview of the Lean-START Model

As highlighted earlier, Lean-START is designed as a tool that allows for rapid computational iteration among the inputs that are used to convey the desired operational capability for a force employment event and the ACS resources required to support mission generation and force beddown. The tool will be described in much greater detail in Chapters 3 and 4. The overview of the tool here is intended to aid in understanding how it might be used in joint and Air Force planning processes described in the next chapter.

For a defined operational mission, the tool produces a list of UTCs needed to support the mission, the sum of the passengers and weight of those UTCs (or footprint), and a computation

---

[12] Don Snyder and Patrick Mills, *Supporting Air and Space Expeditionary Forces: A Methodology for Determining Air Force Deployment Requirements*, Santa Monica, Calif.: RAND Corporation, MG-176-AF, 2004.

[13] Don Snyder, Patrick Mills, Manuel Carrillo, and Adam Resnick, *Supporting Air and Space Expeditionary Forces: Capabilities and Sustainability of Air and Space Expeditionary Forces*, Santa Monica, Calif.: RAND Corporation, MG-303-AF, 2006.

[14] Mahyar Amouzegar, Robert S. Tripp, Ronald G. McGarvey, Ed Chan, and Charles Robert Roll, Jr., *Supporting Air and Space Expeditionary Forces: Analysis of Combat Support Basing Options*, Santa Monica, Calif.: RAND Corporation, MG-261-AF, 2004; Mahyar Amouzegar, Ronald G. McGarvey, Robert S. Tripp, Louis Luangkesorn, Thomas Lang, and Charles Robert Roll, Jr., *Evaluation of Options for Overseas Combat Support Basing*, Santa Monica, Calif.: RAND Corporation, MG-421-AF, 2006.

[15] Brent Thomas, Mahyar Amouzegar, Rachel Costello, Robert A. Guffey, Andrew Karode, Christopher Lynch, Kristin Lynch, Ken Munson, Chad J. R. Ohlandt, Daniel M. Romano, Ricardo Sanchez, Robert S. Tripp, and Joseph Vesely, *Project AIR FORCE Modeling Capabilities for Support of Combat Operations in Denied Environments*, Santa Monica, Calif.: RAND Corporation, RR-427-AF, 2015.

of the number of movement resources (for example, C-17s, C-130s, or trucks) required to transport the UTCs to the mission site.

The operational mission requirements are defined by specifying the numbers and types of weapon systems being employed (for example, F-22, F-15, F-16, or C-130), the daily flying profile planned for those weapon systems (that is, number of sorties per aircraft, duration of the sortie, and type of munitions being flown), the duration of the operation, and the number of days that flying operations will be performed. With the operational mission defined, Lean-START allows the user to evaluate the trade-offs between speed (driven by the total number of people and weight of cargo that must be moved) and the capability to support the flying operations and people being deployed as part of the UTCs.

The capabilities represented in the UTC list generated by Lean-START will generally fall into one of three capability categories: those that directly support the operational flying mission, those that provide safety and security for the aircraft and people, and those that support the care and feeding of the people. For each of these capability categories, the user can specify the level of capability desired, working from a baseline set of capabilities that is computed as a function of the amount of infrastructure available at the beddown location.

The level of capability that the user selects generally affects two factors: the overall amount of people and equipment that must be deployed, and the amount of risk to mission, safety, security, and personnel support being accepted. If, driven by a desire to deploy more quickly by reducing the footprint, the user elects to reduce the capability in one or more areas, they are essentially reducing the ability to mitigate risk to mission execution, aircraft and people, safety and security, or care and feeding of the people.

This type of trade-off analysis may sound intuitive and elementary, but such an automated capability did not exist previously, and planners would spend days or weeks computing and assessing these trade-offs, when the current environment demands the capability to do it much more quickly.

Organization of This Report

The purpose of this report is to describe the basic philosophy and operation of Lean-START, use it to demonstrate basic calculations of lean force packages, and then illuminate some of the basic dynamics and drivers of lean deployments. Chapter 2 describes the joint planning process (which the Air Force supports) and the role that Lean-START could play in supporting planning, including assessing the trade-offs between speed and risk tolerance. Risk is a natural element of the planning process, but it is a particular focus in this planning environment, where speed is crucial and risk management often brings significant transportation demands. Chapter 3 describes the Lean-START model, its input parameters and basic functioning. Chapter 4 shows illustrative calculations using the model. Finally, Chapter 5 offers conclusions and recommendations.

2. The Planning Process and the Role of Lean-START

Organizations within DoD—including combatant commands, the Joint Staff, and the Air Force—conduct planning using a common process and framework. This chapter describes that process and highlights the role of Lean-START analysis in it.

Changing Dynamics of Force Planning

The dynamics of planning force deployments has changed over the years. During the Cold War era, heavy emphasis was placed on well-structured plans against a known enemy, its capabilities, and its strategies. The United States had a significant forward presence on overseas installations that factored into its ability to respond quickly to anticipated enemy aggression. Force deployments were designed to forward-position additional units in response to enemy aggressions. The process of planning for those major contingencies was calculated and deliberate. On occasion, the military would respond to other contingencies worldwide, such as providing humanitarian assistance, and would plan for them using a model for crisis action planning, which was similar to the deliberate planning process but conducted on a compressed timeline.

The environment has changed. The end of the Cold War was followed by the Gulf War and ongoing operations in Southwest Asia. Force structure drawdowns on overseas installations placed a premium on rapid deployment. As highlighted in Chapter 1, the need for rapid deployment resulted in the evolution of the EAF for the Air Force. The shift in force deployment concepts also drove changes to the process by which force deployments were planned.

The Joint Planning Process

The joint planning and execution community (JPEC) operates from a framework for deliberate and crisis action planning called Adaptive Planning and Execution (APEX). Joint Publication 5.0 provides guidance for and characterizes APEX:

> Joint operation planning occurs within Adaptive Planning and Execution (APEX), which is the department-level system of joint policies, processes, procedures, and reporting structures. APEX is supported by communications and information technology that is used by the joint planning and execution community (JPEC) to monitor, plan, and execute mobilization, deployment, employment, sustainment, redeployment, and demobilization activities associated with joint operations. APEX formally integrates the planning activities of the JPEC and facilitates the JFC's [joint force commander's] seamless transition from planning to execution during times of crisis. APEX activities span many organizational levels, but the focus is on the interaction between SecDef [the Secretary of Defense] and CCDRs, which ultimately helps the President and

SecDef decide when, where, and how to commit United States (US) military forces.[16]

A key characteristic of adaptive planning and execution is iterative analysis of potential COAs.[17] The U.S. military no longer has a large forward presence or an abundance of resources from which it can draw to launch military operations. Resources committed to one operation can put other operations around the world at risk. The joint operations planning process, shown in Figure 2.1, appears to take this into consideration and prompts ongoing analysis of COAs to find the most efficient way to conduct effective operations.

The process depicted in Figure 2.1 focuses on planning reviews and analytical iterations within the CCDR's planning staff across the functions, and also between the CCDR and his or her components and their planning staffs. As shown at the top of Figure 2.1, after mission analysis is complete, the process of developing and evaluating various COAs takes place. Each COA that is put forward shapes the operational approach and considers the impact on resources needed to support the COA. The analyses must consider the total force package needed to execute the COA, the speed with which the COA can be executed, and its risk. The gray arrow through the center of the figure indicates that each COA undergoes iterative analysis and is compared with other COAs prior to the final COA selection.

A similar process is used by Air Force components when considering courses of action to support the CCDR.

---

[16] Joint Publication 5–0, *Joint Operational Planning*, U.S. Department of Defense, August 11, 2011, pp. IX–X.

[17] Joint Publication 5–0, 2011.

Figure 2.1. Joint Operational Planning Process

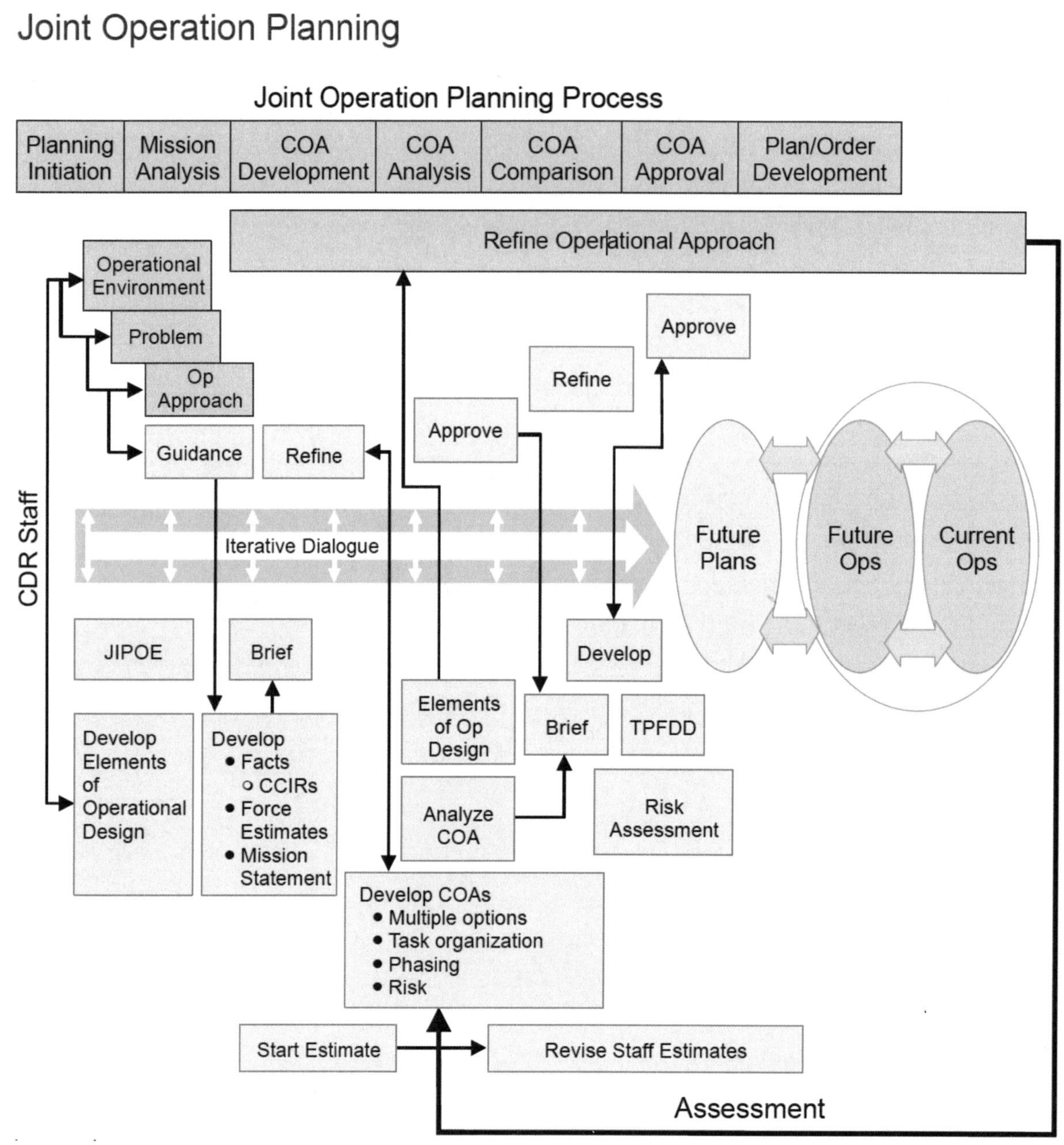

SOURCE: Adapted from Joint Publication 5–0, 2011.
NOTES: CCIR = commander's critical information requirement; JIPOE = joint intelligence preparation of the operational environment; TPFDD = time-phased force and deployment data.

Air Force Planning to Support Joint Operations

The Air Force component to the CCDR receives guidance from him or her that outlines the objectives and roles of Air Force forces in support of the contingency. The Air Force convenes an operational planning team (OPT), typically led by the A5 (Plans) staff with participation from other functions on the A-staff (e.g., A2 [Intelligence] or A4 [Logistics]), and proceeds with developing the Air Force plan.[18] With guidance and performance objectives from the CCDR, the Air Force component will consider the size of the force, types of weapon systems, sortie rates and durations for those weapon systems, and timing and phasing of the force deployment. This collective set of information about the operations, commonly referred to as the *operational demands*, must then be supported by ACS resources for enabling mission operations and supporting the installations that will host the combat forces.

The process of determining support requirements and managing I&MS resources has long been a focus of PAF research, often referred to as combat support planning, execution, and control (CSPEC). PAF's body of research recommends methods for better aligning ACS resources utilization and consumption to operational demands and adjudicating resource allocation when demands exceed available resources.[19] A core element of the research is a process, shown in Figure 2.2, that highlights the link between operational demands and resource supply provided by ACS.

Similar to the joint operations planning process, the notion is that the operational planners (representing the demand) and combat support planners (representing the supply) will iterate on a plan until a course of action that provides the needed operational capability can be met with available resources. This iterative process is depicted by the red circle on the left side of the figure and is similar to the gray arrow depicting the iterative element of the Joint Planning Process in Figure 2.1. Throughout the iterative process, the planners representing both the operations and logistics communities must have an opportunity, informed by analysis, to discuss the trade-offs associated with each COA. What operational forces (demand-side requirements) are needed to execute the CCDR's guidance (light purple box)? What ACS resources (supply-side capabilities) are required to the support the operational forces? How quickly can they be in place at the deployed location? At what performance level must the ACS network be capable of

[18] Discussions with HQ U.S. Air Forces, Europe (USAFE) personnel on December 1–3, 2015 and HQ Pacific Air Forces (PACAF) personnel on July 26–28, 2016.

[19] A series of PAF reports on C2 and enhanced ACS processes highlighted the value of improving the links between logisticians and warfighters with the intent of guiding resource allocation in a manner that improves operational capability and effectiveness. The area of research has also been conducted under different terminology that includes ACS C2 and Combat Support Planning, Execution, Monitoring and Control. See James A. Leftwich, Robert S. Tripp, Amanda B. Geller, Patrick Mills, Tom LaTourrette, Charles Robert Roll Jr., Cauley von Hoffman, and David Johansen, *Supporting Expeditionary Aerospace Forces: An Operational Architecture for Combat Support Execution Planning and Control*, Santa Monica, Calif.: RAND Corporation, MR-1536-AF, 2002; Robert S. Tripp, Kristin F. Lynch, John G. Drew, and Robert G. DeFeo, *Improving Air Force Command and Control Through Enhanced Agile Combat Support Planning, Execution, Monitoring, and Control Processes*, Santa Monica, Calif.: RAND Corporation, MG-1070-AF, 2012; Kristin F. Lynch, John G. Drew, Robert S. Tripp, Daniel M. Romano, Jin Woo Yi, Amy L. Maletic, *Implementation Actions for Improving Air Force Command and Control Through Enhanced Agile Combat Support Planning, Execution, Monitoring, and Control Processes*, Santa Monica, Calif.: RAND Corporation, RR-259-AF, 2014a; Kristin F. Lynch, John G. Drew, Robert S. Tripp, Daniel M. Romano, Jin Woo Yi, Amy L. Maletic, *An Operational Architecture for Improving Air Force Command and Control Through Enhanced Agile Combat Support Planning, Execution, Monitoring, and Control Processes*, Santa Monica, Calif.: RAND Corporation, RR-261-AF, 2014b.

operating to sustain the operational mission (gray assessment box)? How will the effectiveness of the operational and ACS systems be measured (red measure-of-effectiveness boxes)?

Figure 2.2. Lean-START Informs COA Development and Analysis

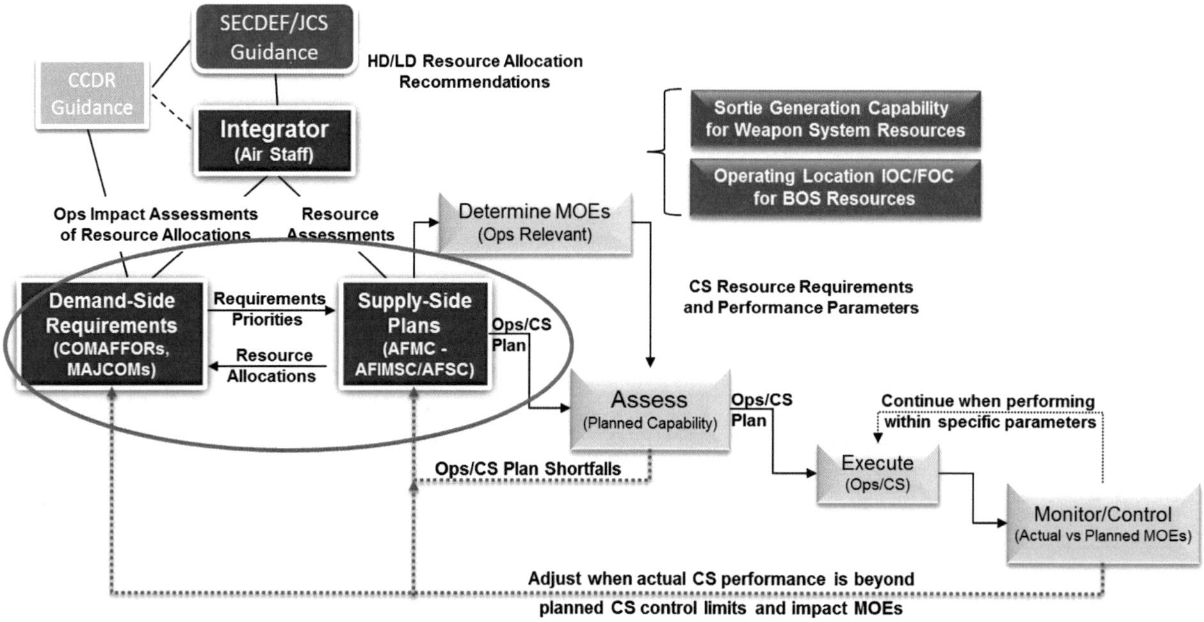

SOURCES: Adapted from Leftwich et al., 2002; and Tripp et al., 2012.
NOTES: This figure depicting ACS planning in conjunction with operations planning appears in a number of different forms in the ACS C2 documents already referenced. JCS = Joint Chiefs of Staff; HD/LD = high-demand/low-density; Ops = operations; COMAFFOR = Commander Air Force forces; MAJCOM = major command; AFMC = Air Force Materiel Command; AFIMSC/AFSC = Air Force Installation and Mission Support Center/Air Force Sustainment Center; MOE = measure of effectiveness; CS = combat support; IOC/FOC = initial operating capability/full operating capability; BOS = base operating support.

By design, each potential COA will have different strengths and weaknesses. Often, the strengths and weaknesses are interrelated across the different COAs, and those interrelationships sit at the nexus of the tradespace that must be considered by decisionmakers. The result of the iterative part of the process is a COA that meets the operational requirements within the logistics constraints or capabilities, with known risks and expected performance outcomes. As the plan is executed, the planning community will continue to monitor the execution and make adjustments as needed to ensure that mission requirements are met. Lean-START is designed to inform the decisions during COA iteration and provide expected performance parameters relative to speed and capability.

Elements of the Tradespace and Lean-START

As described earlier, the CCDR's guidance to components will describe objectives that the Air Force must meet with airpower resources to support the overall strategy. Those objectives equate to the need for different weapon systems, performing at various operation tempos, within certain timelines. Those operational parameters subsequently equate to the need for combat support resources to support generation of the operational mission and also support the beddown, safety, and security of the force at a forward-deployed installation. All of these elements factor into a tradespace that includes size and capability of the operational force, level of support that will be available at the beddown location, and speed at which the operations and support forces can be deployed.

A force capable of delivering lower operational effects can generally be deployed faster. The differences in capability could be a function of the number of aircraft, the type of aircraft (e.g., fourth generation versus fifth generation),[20] the types of missions that the aircraft will be expected to fly (e.g., air-to-air versus air-to-ground), or even the expected sortie rates and durations for the aircraft. Generally speaking, a force package with a smaller number of fourth-generation aircraft flying a low sortie rate and capable of carrying only air-to-air munitions is much more responsive, but less capable, than the same number of fifth-generation aircraft flying higher sortie rates with the full range of munitions.

When the operational objectives are such that a specific weapon system performing at a certain level is essential to successful outcomes of the campaign, planners are forced to consider other ways of reducing the footprint to enable a rapid response. That is where risk becomes an element of the tradespace.

Articulating and managing risk is a part of the COA iterative process for both the joint community and the Air Force. Joint policy highlights this:[21]

> Assessing risk and identifying mitigation strategies are fundamental to joint operation planning. In the course of developing multiple options to meet the strategic end state, JFCs and their planning staffs, as well as the larger JPEC, identify and communicate shortfalls in DoD's ability to resource, execute, and sustain the military operations contained in the plan as well as the necessary actions to reduce, control, or accept risk with knowledge of potential consequences. JFCs communicate risk to senior leadership during in-progress reviews (IPRs) of the plan.

One way to increase the responsiveness of the deploying force is by reducing the size, or number of short tons, of the total package that must be deployed. A less-capable force is generally a lighter, more responsive force. The size of the deployment package can also be reduced by accepting more risk to the force—to reduce the package, equipment and personnel must be removed that provide a capability at the beddown site tied to the security, safety, or care and feeding of the force. The willingness of a commander to accept a certain level of risk can directly affect the speed at which the force can be deployed.

---

[20] Aircraft technology adoption and production cycles usually occur over several decades. As such, they are generally referred to in generations that group together particular production and technology capability cycles. The term *fourth generation* is used to refer to such aircraft as the F-16 and F-15, while the term *fifth generation* is used to refer to such aircraft as the F-22 and the F-35.

[21] Joint Publication 5–0, 2011, p. I–2.

In the context of planning and assessing various COAs, analyzing risk is accomplished under the COA analysis metric of "acceptability." When considering acceptability, Joint Publication 5.0 states that the analysis "must balance cost and risk with the advantage gained."[22] It goes on to suggest the types of questions that must be considered as part of the acceptability analysis:

> Are COAs reconciled with external constraints, particularly rules of engagement (ROE)? This requires visualization of execution of the COA against each enemy capability. Although this process occurs during COA analysis and the test at this time is preliminary, it may be possible to declare a COA unacceptable if it violates the commander's definition of acceptable risk.[23]

The Air Force policy describes a framework for assessing risk related to the security of a beddown location. The process shown in Figure 2.3 considers the importance of the asset being protected, the threat to the asset, and the vulnerability of the asset to the threat. Lean-START is designed as an analytical capability to inform decisions in this tradespace and do it within a time frame that supports the iterative element of both the joint operations planning process and the Air Force's planning process. Specific to this process, Lean-START helps classify the criticality of the asset relative to a risk, and it helps support the cost-benefit analysis that eventually informs the commander's decision whether to accept the risk.

Figure 2.3. Air Force Process for Risk Assessment

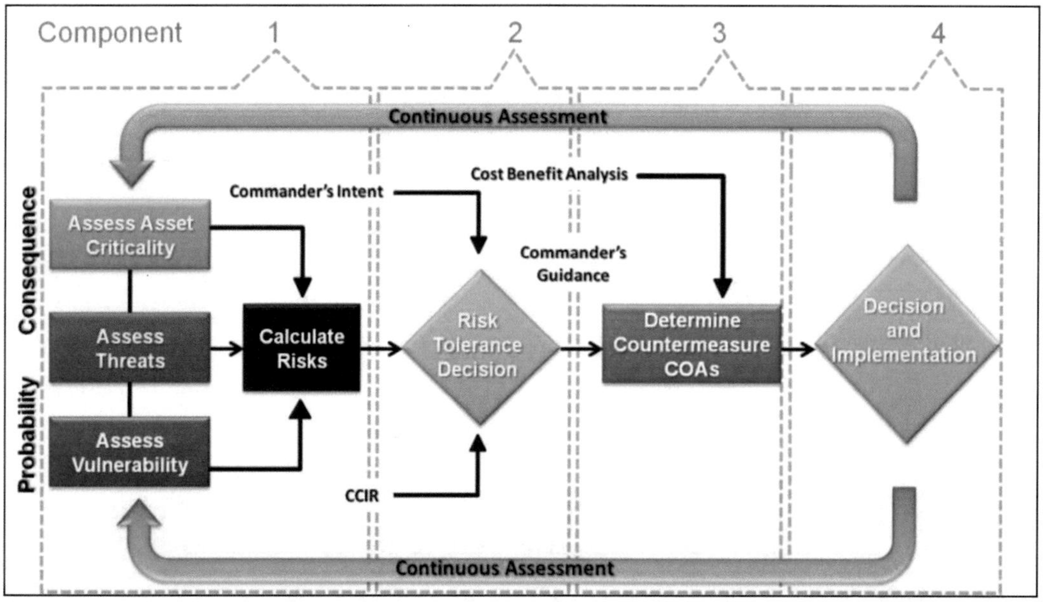

SOURCE: Air Force Instruction 10–245, *Antiterrorism (AT)*, U.S. Air Force, June 25, 2015.

---

[22] Joint Publication 5–0, 2011, p IV–24.

[23] Joint Publication 5–0, 2011, p. IV–25.

Articulating and Managing Risk in Deployment Planning

ACS planners and practitioners are naturally conservative or risk-averse: They seek to ensure mission success and keep Air Force personnel and property safe and secure. The Air Force has an array of policies and procedures for risk acceptance and management. Some of them involve constraints on activities (e.g., mixing dangerous types of cargo or munitions); some, safety procedures (especially for aircraft maintenance); and others, the resources required to adequately perform a task (e.g., firefighting vehicles and crews to extinguish an aircraft fire). The appendix of this report provides additional detail on the range of Air Force risk management policies.

In our discussion of risk in the context of deployed operations, mitigating or decreasing risk often requires increasing (or maintaining) resource requirements. Traditional Air Force planning factors and assumptions (what we call the "baseline") are generally risk-averse and have risk management built into them. The deployment planning process further explores the tension between speed and risk to arrive at feasible COAs, often asking planners to "do with less" than they would naturally plan to.

Impacts of Resource Limitations on Deployments

Resource limitations placed on ACS functional areas can have very different impacts on deployed operations. We start by broadly describing the nature of those impacts, with the aim of targeting the most critical types of risk.

Joint policy focuses on risk more generally in terms of the ability to accomplish the mission; however, we further define risk relative to the different mission areas that can be affected. We group mission areas and the risk effects of resource shortfalls into four categories.

- **Mission capacity and duration.** By this, we mean the ability to reliably generate effective sorties. Ultimately, every person and equipment item at a contingency operating location in some way supports sortie generation, but here we include those that have a direct and quantifiable impact.
- **Safety.** We define this as protection from the occurrence or risk of injury, danger, or loss due to *internal* causes. Many of the Air Force's formalized risk management policies and practices target this category.
- **Security.** Security is often defined in a way to overlap with safety. In this context, we mean protection against *external* threats, including attack, sabotage, and espionage. This is almost the exclusive purview of security forces (SF).
- **Human performance and endurance.** Providing care and provisions to deployed forces affects the health, well-being, and morale of deployed forces. The quality of care and provisions can affect productivity and thus put mission success at risk. This category focuses on such items as living quarters and quality of subsistence (e.g., meals ready to eat as opposed to deployable dining facilities).

We now apply those four categories to the functional capabilities in Lean-START. Table 2.1 shows the relationship between functional area and the type of impacts. An "X" indicates that taking less than required resources for that functional area may have an effect on that category.

Table 2.1. Functional Area Impacts of Resource Shortfalls

Functional Area	Mission Capacity and Duration	Safety	Security	Human Performance and Endurance
Aircraft maintenance	X			
Munitions maintenance	X	X		
Fuels	X			
Munitions	X	X		
Communications	X			X
Aerial port operations	X			
CBRN		X		
Explosives ordnance disposal	X	X		
Fire services		X		
Medical		X		X
Security forces			X	
Prime BEEF	X	X		X
BEAR	X	X	X	X
Vehicles	X	X	X	X
Services				X

NOTES: CBRN = chemical, biological, radiological, nuclear; BEAR = basic expeditionary airfield resources; Prime BEEF = Prime Base Engineer Emergency Force.

In Chapter 4, we apply these definitions in example calculations to show how much of the footprint for different deployments applies to which category, which helps us better understand the effects of reducing resources in the interest of speeding deployment times.

To increase speed of deployment, the heavier items should be considered for exclusion in the deployed force package, but commanders must look at the level of risk they are accepting with such exclusions. We again turn to Joint Publication 5.0 and how it addresses risk considerations:

> Determining the risk is more an art than a science. Planners use historical data, intuitive analysis, and judgment to estimate the risk of each threat. Probability and severity levels are estimated based on the user's knowledge of probability of occurrence and the severity of consequences once the occurrence happens. The level of risk is assigned by assessing the hazards'/obstacles' probability of occurring and their degree of severity. The levels of risk are:
>
> (a) Extremely high: loss of ability to accomplish mission;
>
> (b) High: significantly degrades mission capabilities in terms of required mission standards;
>
> (c) Moderate: degrades mission capabilities in terms of required mission standards; and

(d) Low: little or no impact on accomplishment of the mission.[24]

Referring back to Table 2.1, we highlight which areas affect mission capacity and duration, and which affect safety, security, and human performance and endurance. When we assessed our functional area impacts of resource shortfall categories against the joint definition for levels of risk, a shortfall in mission capacity and duration could reasonably be categorized as an extremely high risk that would result in the loss of ability to meet mission requirements. The other three categories for the effect of resource shortfalls could be examined from the degree to which they affect "mission standards," which fall into the high, moderate, and low levels of risk. Accordingly, a commander might be more inclined to look at safety, security, and human performance and endurance when accepting risk to increase speed.

Lean-START and Risk

As we highlighted earlier referencing Joint Publication 5.0, risk trade-off analysis is a core element of adaptive planning and execution. While the CCDR's planning team is considering various courses of action, their planning must be informed by planning being conducted by the Air Force. As we learned from discussions with planners in the Pacific and European theaters, the frequency of planning and speed at which it must be accomplished drives the need to rapidly consider various courses of action and the strengths and weaknesses of each. The iterative planning process must be informed by meaningful analysis that allows for costs and benefits of risk acceptance to be quantified and communicated in order to inform decisions.

In military operations, it is difficult to fully quantify risk (including likelihood, not just effect); as we pointed out earlier, most risks are assessed and broadly classified as extremely high, high, moderate, or low. On the other hand, the benefits of risk acceptance can be quantified relative to their value in improving speed of deployment and force agility. Informing this risk trade-off dialogue is a fundamental capability of Lean-START. The model can quickly quantify how much "speed of deployment" can be gained by eliminating certain equipment or personnel and accepting the risk associated with the reduction of those resources. We examine the specifics of the risk/speed trade-offs in more detail in Chapter 4 using the Lean-START model.

---

[24] Joint Publication 5–0, 2011, p. IV–11.

3. How the Model Works

This chapter describes the basic workings of the Lean-START model. We describe the methodology and processes used to develop the tool, the basic flow and user inputs for the model, and the types of outputs that are available.

Data Sources and Rule Sets in Lean-START

The most important contribution of Lean-START is its deployment rule sets. When we say *deployment rules* or *deployment logic* (used interchangeably in this report), we simply mean if-then statements that tie desired capabilities or existing conditions at an operating location to one or more UTCs that contain the personnel and equipment necessary to provide the needed capability under the user-specified conditions. Table 3.1 shows some examples of this if-then logic from the model.

Table 3.1. If-Then Statements Typical of Lean-START Rule Sets

Condition ("If")	Requirement ("Then")
12 x F-16 aircraft deploy	Send aviation package containing pilots and life support equipment (UTC 3FKS1)
UTC 3FKS1 deploys	Send HFKS1 maintenance and HGKS1 munitions maintenance packages
CBRN threat is "High"	Send Prime BEEF EM 4F9W-series UTCs consisting of 18 personnel and 47 tons of equipment
Base population is between 1,500 and 3,000	Send the EMEDS Basic medical package

NOTES: Each UTC is represented by a five-character alphanumeric code (e.g., 3FKS1, HFKS1). EOD = explosives ordnance disposal; EM = emergency management; EMEDS = expeditionary medical deployment system.

Figure 3.1 shows the sources for Lean-START and how they inform the tool. We used four groups of inputs, depicted on the left side of the figure.

Figure 3.1. Lean-START Sources

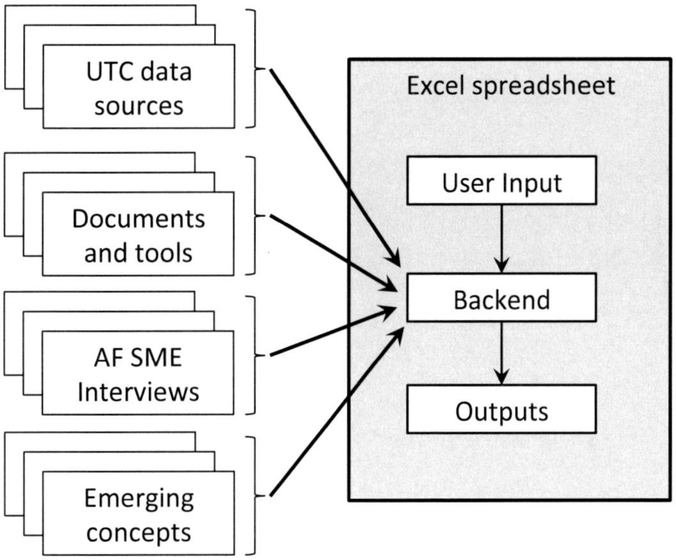

Data Sources to Inform Lean-START

First, we drew from several UTC data sources to obtain the official data the Air Force uses to express its deployable capabilities and their movement requirements. This includes the manpower and equipment force package (MEFPAK), mission capability (MISCAP) statements, logistics detail (LOGDET), and manpower force package system (MANFOR).

The MEFPAK is a list of UTCs with high-level movement characteristics (manpower in the form of authorized personnel and passengers, and equipment in the form of short tons in different categories). MEFPAK data can be married with more-detailed manpower and logistics details to obtain the full details of their composition. Likewise, each UTC has a MISCAP statement that explains in narrative form the task to be performed and its conditions. For example, UTC 3FKP1 provides an aviation package for 12 F-16 aircraft (e.g., pilots, flight-line maintenance):

> ...PROVIDES IND 12 SHIP FTR SPT FOR CONTINGENCIES AND/OR GENERAL WAR. SORTIE DURATION AND EXPENDITURE RATES ARE IAW [in accordance with] WMP-5 [War Mobilization Plan, Vol. 5]. CAN OPERATE UP TO 30 DAYS AT A FOB FLYING WMP-5 RATES BASED ON MRSP. EMPLOYS CONVENTIONAL MUNITIONS DAY/NIGHT (INCLUDING PGMS LISTED ON UCML) IN THE SA, AI, OCA-S, SEAD, DEAD, OCA-A AND DCA ROLES . . .

The corresponding maintenance package for that is contained in UTC HFKP1. Its MISCAP statement says:

> SUPPORTS UTC 3FKP1. PROVIDES INITIAL MAINTENANCE SUPPORT AND AN INDEPENDENT MRSP FOR 12 F-16 BLK 50 AIRCRAFT. CAPABLE OF BARE BASE OPERATIONS . . .

A plethora of these interrelationships exist, and we captured those that were appropriate to the scope of the model, particularly for aircraft and their supporting maintenance packages. When linked, these data sets provide two things: a starting point for the rule sets, and detailed manpower and equipment information needed to parse Air Force capabilities. The integration of these data sets forms the backbone of the model.

Air Force Documentation and Tools to Inform Lean-START

Second, we drew from official and unofficial Air Force documentation and tools. Some functional areas have official Air Force publications (e.g., instructions, manuals, pamphlets) that are periodically updated to distribute planning factors. Others have less official sources: For example, the *Civil Engineer Supplement to the War and Mobilization Plan-1 (WMP-1)* has conditions under which various UTCs deploy, and their associated levels of capabilities.[25]

Figure 3.2 shows screen shots from the WMP-1 supplement. The two panels show Prime BEEF UTCs, which include a range of skill sets that support many basic engineering functions on base, such as planning and surveying, electrical power, water and wastewater, carpentry, and pest management. These UTC requirements are population-driven. These two panels show rules for base populations of 550 and 1,100 personnel—i.e., the number of each UTC that should deploy in each case. The guidance accommodates populations of up to 3,300 personnel. We adopted these rules wholesale.

[25] U.S. Air Force, *Civil Engineer Supplement to the War and Mobilization Plan-1 (WMP-1) 2016*, Washington, D.C., 2016.

Figure 3.2. Illustrative Prime BEEF Rule Sets from WMP-1 Supplement

550 person beddown

Qty	UTC	Title	Qty	UTC	Title
Prime BEEF Engineering/Operations Personnel (66-pn)			**Prime BEEF Engineering and Operations Equipment**		
2	4FPET	EN Basic Eng Beddown/Sustainment Team	1	4F9ET	EN Basic Eng Beddown/Sustainment Eq Set
1	4FPES	EN Command and Control	1	4F9EF	EN Beddown/Sustain Follow-on Eq Set
2	4FPAN	EN Operations Superintendent	1	4F9ER	EN Beddown/Sustain Comm Lead Eq Set
1	4FPAX	EN Pest Management Team	1	4F9EH	EN Beddown/Sustainment Survey Supt Eq Set
1	4FPAM	EN Logistics Support Team	1	4F9EE	EN Beddown/Sustainment Pest Mgt Supt Eq
1	4FPAP	EN Power Generation/AAS Team	1	4F9S6	EN Command and Control Support Eq
For Prime BEEF CBRN and EM requirements see Attachment 10 FUNCTIONAL FORCE MODULE—Prime BEEF CBRN and EM Teams					
For Prime BEEF FES aircraft support see Attachment 9 FUNCTIONAL FORCE MODULE—Prime BEEF FES Teams					
For Prime BEEF EOD requirements see Attachment 13 FUNCTIONAL FORCE MODULE—Prime BEEF EOD Teams					

1100 person beddown

Qty	UTC	Title	Qty	UTC	Title
Prime BEEF Engineering/Operations Personnel (102-pn)			**Prime BEEF Engineering and Operations Equipment**		
1	4FPES	EN Command and Control	1	4F9ET	EN Basic Eng Beddown/Sustainment Eq Set
3	4FPET	EN Basic Eng Beddown/Sustainment Team	2	4F9EF	EN Beddown/Sustain Follow-on Eq Set
1	4FPAL	EN Fuel Systems Maintenance Team	1	4F9ER	EN Beddown/Sustain Comm Lead Eq Set
2	4FPAM	EN Logistics Support Team	1	4F9ED	EN Beddown/Sustain Follow-on Comm Eq
3	4FPAN	EN Operations Superintendent	1	4F9EH	EN Beddown/Sustainment Survey Supt Eq Set
1	4FPAP	EN Power Generation/AAS Team	1	4F9EE	EN Beddown/Sustainment Pest Mgt Supt Eq
1	4FPAR	EN Heat, A/C, Refrigeration & Control Team	1	4F9AL	EN Fuels System Maintenance Eq Support Set
1	4FPAX	EN Pest Management Team	1	4F9S6	EN Command and Control Support Eq
1	4FPSB	EN Operations Field Grade Officer			
1	4FPSD	EN Engineer Company Grade Officer			
For Prime BEEF CBRN and EM requirements see Attachment 10 FUNCTIONAL FORCE MODULE—Prime BEEF CBRN and EM Teams					
For Prime BEEF FES aircraft support see Attachment 9 FUNCTIONAL FORCE MODULE—Prime BEEF FES Teams					
For Prime BEEF EOD requirements see Attachment 13 FUNCTIONAL FORCE MODULE—Prime BEEF EOD Teams					

SOURCE: U.S. Air Force, 2016.

Figure 3.3 shows similar rules, but for firefighting. Here, the requirements are driven by the size of the aircraft (so that the combination of personnel and equipment can extinguish an aircraft fire in a sufficient time period). The Air Force uses National Fire Protection Association (NFPA) standards for its baseline planning factors. The rules pictured here are for NFPA airport categories 6/7 and 8 (they range from 6 to 10). The tables have explicit rules for personnel and equipment UTCs driven by each type of aircraft. We also adopt these rules as-is for our baseline planning factors (i.e., robust, sustained capability). We will discuss how we depart from these baseline firefighting rules for lean deployments.

Figure 3.3. Illustrative Firefighting Rule Sets from WMP-1 Supplement

Category 6/7 Aircraft – 5,000 gallons of firefighting agent: (A-10, AC-130, B-1, C-9, C-12, C-20, C-21, C-27, C-22, C-32, C-37, C-38, CV-22, C-130, 707, 737, MD-80, MC-130, MH-53, F-15, F-16, F-22, FB-111, F-117, T-1, T-6A, T-37, T-38, T-39, T-43, RQ-1A/B, RQ-4A, rotary wing not in other categories)

Qty	UTC	Title	Qty	UTC	Title
Prime BEEF FES Personnel (54-pn) (See Note)			**Prime BEEF FES Equipment**		
1	4FPFF	EN Fire Chief Manager	1	4F9FE	Firefighter Communications Package
2	4FPFJ	EN Firefighter Management 2 Pk Team	1	4F9FF	Firefighter SCBA Compressor
1	4FPFN	EN Deputy or Fire Chief Manager	1	4F9FR	Fire Night Vision Goggles
8	4FPFP	Firefighter Truck Crew	1	4F9FX	Firefighter Limited Eq Set
			1	UFM33	P-26 Water Tanker Firefighting Vehicle
			1	UFM34	P-24 Pumper Firefighting Vehicle
			5	UFM38	P-19 Crash Rescue Firefighting Vehicle
			1	UFM3A	P-10 Rescue Vehicle

Category 8 Aircraft– 8,000 gallons of firefighting agent: (B-2, B-52, C-17, KC-135)

Qty	UTC	Title	Qty	UTC	Title
Prime BEEF FES Personnel (60-pn) (See Note)			**Prime BEEF FES Equipment**		
1	4FPFF	EN Fire Chief Manager	1	4F9FE	Firefighter Communications Package
2	4FPFJ	EN Firefighter Management 2 Pk Team	1	4F9FF	Firefighter SCBA Compressor
1	4FPFN	EN Deputy or Fire Chief Manager	1	4F9FR	Fire Night Vision Goggles
9	4FPFP	Firefighter Truck Crew	1	4F9FX	Firefighter Limited Eq Set
			1	UFM33	P-26 Water Tanker Firefighting Vehicle
			1	UFM34	P-24 Pumper Firefighting Vehicle
			8	UFM38	P-19 Crash Rescue Firefighting Vehicle
			1	UFM3A	P-10 Rescue Vehicle

SOURCE: U.S. Air Force, 2016.
NOTES: Qty = Quantity; SCBA = Self-Contained Breathing Apparatus.

Likewise, several areas have decision support tools specific to their area. BEAR and fuels are good examples of this.[26] Others simply have content in spreadsheets or briefings. Many SMEs provided spreadsheets with lookup tables (similar to the ones in the WMP-1) they use to do deployment planning but are not published. We drew as much information as we could from these sources.

Table 3.2 lists many, though not all, of the documents and tools we used to develop our rule sets.

[26] We found the BEAR tool to have too much detail to utilize for our purposes. For that single functional area, the tool had more than a dozen inputs, most of which are not directly or indirectly obtainable from the more minimalistic Lean-START inputs. The fuels equipment tool was quite useful. We analyzed its rule sets, adapted the equipment items to match the latest in the inventory, and simplified some of the mathematics, but largely adopted its rule sets as given.

Table 3.2. Sources for Lean-START Rule Sets

Functional Area	Source	Primary Driving Factors
Aircraft support[a]	MEFPAK, MISCAP	Type and number of aircraft
Civil engineering[b]	WMP-1 supplement	Population, threat
Medical	Spreadsheets provided by SMEs	Population
Communications	Spreadsheets provided by SMEs	Population
Services	Spreadsheets provided by SMEs	Population
Fuels support	Fuels Support Equipment (FSE) Calculator, spreadsheets provided by SMEs	Type and number of aircraft, fuel consumption and storage
BEAR	Spreadsheets provided by SMEs	Population, type of aircraft
Airfield damage repair (ADR)	Spreadsheets provided by SMEs	ADR package size (e.g., small, large)

[a] Includes aviation, maintenance, and munitions maintenance.
[b] Includes Prime BEEF, firefighting, EOD, and CBRN.

Nearly every functional area had a document (official or unofficial), a tool, or an SME spreadsheet we could adopt or adapt. SF is the major exception. Parameterizing SF requirements is notoriously difficult because the threats can vary so much, so we depended more on SME interviews.

Interviews with Subject Matter Experts to Inform Lean-START

The sources listed in Table 3.2 cover most ACS functions on Lean-START and formed the backbone of our baseline rule sets. But the point of Lean-START is to go beyond these standard rule sets—which are predominantly designed for robust, sustained presence—to provide more-granular requirements for deployments that are smaller, shorter duration, or in some way accept less-than-standard capability.[27] Thus, we conducted focused interviews with a range of Air Force SMEs from headquarters (HQ) PACAF, HQ USAFE, HQ Air Combat Command (ACC), and HQ Air Mobility Command (AMC) (depicted as the third input on Figure 3.1). These meetings helped us clarify ambiguities in documentation and tools and dig deeper into deployment logic. In particular, because we were often decomposing UTCs into their component parts, we had to understand what drove the deployment of individual skill sets and equipment types, something rarely contained in the documentation itself.

Emerging Concepts to Inform Lean-START

Fourth, we drew on emerging concepts being explored by various groups in the Air Force, including but not limited to Rapid Raptor, Rapid-X, Untethered Operations, and Adaptive Basing. All of those concepts have elements intended to increase the survivability of combat air forces (particularly fighters) by disaggregating operational units into smaller components,

---

[27] For example, standard aircraft maintenance UTCs are sized and shaped to meet the most demanding WMP-5 requirements for full-spectrum operations and maximum sortie generating capacity.

decoupling them from their maintenance support, and operating the flying units in concert with a network of bases with various levels of aircraft servicing capability, all of which contrasts with the current way of deploying fighter units to fixed bases with robust maintenance support, generating a high volume of sorties over a sustained time period.

The Rapid Raptor concept seeks to quickly deploy a package of F-22 Raptors and supporting logistics to any forward operating base in the world, and have the aircraft in combat-ready status within 24 hours of employment. The deployment package uses at least one C-17 aircraft carrying supporting materials, munitions, and maintainers to swiftly move, refuel, and rearm a minimum of four F-22s in unfamiliar, austere environments, with a small footprint.[28] Rapid Raptor appears to be a PACAF concept originally developed by the F-22 unit at the 3rd Wing at Joint Base Elmendorf-Richardson, Alaska.

Rapid-X, very similar to Rapid Raptor but developed by HQ USAFE, calls for the deployment of fighter aircraft (not just F-22s) in an agile and quick way to accommodate missions from bases that do not necessarily have the full infrastructure that usually accompanies fighter units in a major contingency. The concept calls, for example, for bringing in four aircraft, rearming and maintaining them, then having them go fly another mission, after which the base might be vacated. The intent is to enable the use of a wide range of locations that are available in Europe to create challenges for a potential adversary that sought to target aircraft and disrupt operations.[29]

Rapid-X appears closely related to untethered operations, another concept that originated in USAFE. This concept seeks to leverage robust basing and North Atlantic Treaty Organization (NATO) partner interoperability to complicate Russian targeting and create an arsenal of options for allied combat operations in Europe. With this concept, a small package of fighters could drop into a base with the support of the personnel and equipment that can fit on a single C-17, conduct operations for several hours without bedding down overnight, then depart again, leaving the base essentially as they found it.[30]

Adaptive Basing is an evolving concept that is broad in scope, focusing on the operational art of arranging the vast array of basing and ACS processes and capabilities to survive and fight in high-threat environments. Adaptive basing proposes alternate basing options to enable flying operations. It calls for forces to disaggregate capabilities from a single base and disperse forces and capabilities to many locations for operational maneuver.[31]

---

[28] "Rapid Raptor Package," *Air Force Magazine*, Sept. 27, 2013.

[29] Jon Harper, "U.S. Air Force Preparing for 'High Volume' Operations in Europe," *National Defense Magazine*, April 5, 2016.

[30] Charles Q. Brown, Jr., Bradley D. Spacy, and Charles G. Glover III, "Untethered Operations Rapid Mobility and Forward Basing Are Keys to Airpower's Success in the Antiaccess/Area-Denial Environment," *Air & Space Power Journal*, May–June 2015.

[31] David Dammeier, Meka Toliver, and Logan Smith, "Future Concepts: Overcoming a Power Projection Problem," *Air Force Civil Engineering Center,* website, undated.

One of the distinctive features of adaptive basing is the expectation that at least part of a theater network of bases will be resourced and operated to provide flexible, responsive support to combat air forces, which may operate for brief periods of time out of one base or another and periodically move from base to base. This could be supported in part by treating these bases more like the global en-route support structure that supports mobility aircraft that fly in and out periodically, are serviced by airmen and equipment not belonging to their home-station unit, and depart again on another mission without remaining overnight or receiving intensive repair actions.

Various groups within the Air Force have been fleshing out these concepts for several years, including the deployment logic that ties the operational force packages to their very lean ACS footprint. We often used these to calibrate our rule sets.

Table 3.3 shows one example of how we went beyond standard deployment rules such as those referenced in Table 3.2 by applying insights from SME interviews and emerging concepts. The first column in Table 3.3 mirrors the firefighting capability selections contained in Lean-START (explained in the next section). For the maximum level of capability—"protect aircraft and structures"—we apply standard Air Force civil engineering (CE) UTC rules. This results in a manpower requirement of between 54 and 66 personnel and eight to 13 vehicles, depending on the NFPA category (which is in turn driven by the type of aircraft).

Table 3.3. Firefighting Deployment Rules in Lean-START

Firefighting Capability Selection	Deployment Rule	Manpower	Vehicles
Protect aircraft and structures	Follow standard Air Force CE UTC rules	54–66	8–13
Protect aircraft only	Provide capability to extinguish aircraft fire and extract pilot	27	3
Extract pilot only	Provide capability to extract pilot but not save aircraft	7	1
C2 only	Provide capability to coordinate with host-nation or other firefighting forces; no organic firefighting capability provided	1	0
No support	No support provided	0	0

Lean-START then has four capability choices below that standard level. Table 3.4 shows the deployment rule (i.e., the type of capability provided), and the resulting manpower and vehicle requirement for each. This allows a user to view the requirement for a standard force package as the default, then down-select the capability level as a means to reduce the footprint requirement.

Several other areas follow this pattern in the model. EOD and CBRN both have the option to allow the user-input threat level to drive the demand, following standard CE planning factors. Or, one can select other options that dial down the capability, all the way to zero if desired.

Rule Sets in Lean-START

We synthesized these sources to develop a complete rule set, and integrated the data sets to produce the back end of the tool. Collectively, we viewed these inputs to represent the baseline set of approved planning factors that the Air Force uses to plan and execute combat operations and associated support operations.

Operation of Lean-START

Model Flow

Figure 3.4 shows the high-level flow of the model. We now briefly describe each step. The subsequent section explains each one in more detail.

Figure 3.4. Lean-START Flow

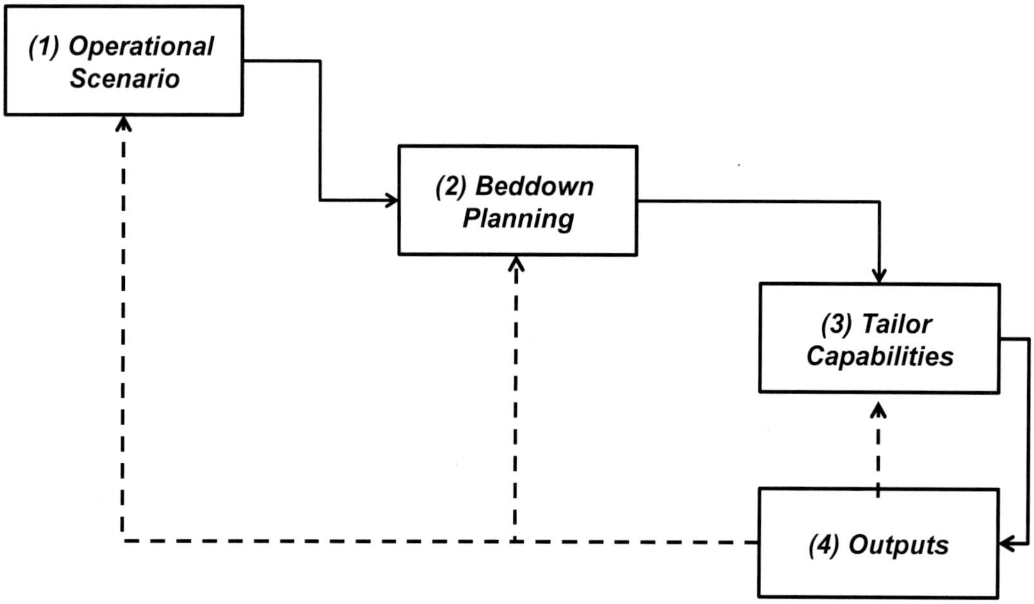

The first input screen is the operational scenario: duration, type and number of aircraft, and more detailed mission information to include sortie rates and mission types.

These inputs drive demands for several areas:

- aviation: pilots and other mission duties
- maintenance: flight-line and repair personnel working directly with the aircraft, its components, and aerospace ground equipment (AGE)
- munitions maintenance: personnel to account for, build up, and transport munitions (chaff, flare, missiles, and bombs) to the flight line
- munitions: the actual weapons themselves.

For beddown planning, there is a screen that has a range of inputs for nonmaintenance ACS. These provide the user input options for both the inherent characteristics of the destination location and the desired level of capability for the operation.

There are multiple screens with which the user can tailor the requirements, given information or assumptions about existing levels of support provided by prepositioned equipment, host-nation support, contract support, or risk tolerance.

Outputs include detailed lists of personnel, equipment, and UTCs on a number of screens, as well as aggregate movement characteristics in the form of personnel, equipment, and transportation requirements. We do not directly address timing in Lean-START—neither the time to deploy from one location to another (e.g., a dispersal scenario) nor the setup time at a new location. In this report, we use transportation requirements as a proxy for time, such that more C-17 loads (or whatever the mode) requires more time to deploy and set up.

Finally, given those outputs, the user can iterate to change inputs to explore the tradespace of options. We now describe each set of inputs in more detail.

Operational Scenario

Operational scenario consists of several inputs, summarized in Table 3.4.

Table 3.4. Operational Scenario Inputs

Duration	Weapon System and Flying Profile
Total duration	Mission Design Series (MDS)
Flying days	Number
Maintenance capability	Sortie rate
	Average sortie duration (ASD)
	Mission type

The *total duration* of the operation drives care and feeding support requirements, such as billeting, food, water, etc. A sustained operation will have an extensive care and feeding footprint; a shorter one could have virtually no support. *Flying days* apply to aircraft operations and drive the maintenance footprint, munitions, and fuel. Maintenance *capability* includes levels of capability from basic flight-line servicing up to full-scale backshop repair. Table 3.5 provides deeper explanation of how the model interprets and operationalizes the combination of "flying days" and "maintenance capability."

Table 3.5. How Lean-START Operationalizes Flying Days and Maintenance Capability

Maintenance Capability Choices	Flying Days Input		
	1–7 Days	8–30 Days	30+ Days
Transient alert	En-route support team (EST)	Same as 1–7 days	Same as 8–30 days
Launch and recover	EST + ADVON team	Same as 1–7 days	Same as 8–30 days
Repair	EST + ADVON + basic remove and replace repair capability	Adds wheel and tire buildup, nondestructive inspection (NDI)	Adds phase maintenance, AGE inspection, intermediate-level repair as applicable

NOTES: ADVON= advanced echelon; CTK = consolidated tool kits; NDI = nondestructive inspection.

In Table 3.5, the left column shows the input range values for "maintenance capability" in Lean-START. The second row shows the input ranges for "flying days." The values in the table cells show the type of capability included for each combination of input parameters.

Transient alert maintenance provides the minimum capability to provide a flyable aircraft. It handles parking, provides power to aircraft, provides refueling, checks and services fluids, troubleshoots and fixes basic maintenance problems, and prepares aircraft for departure. For any number of flying days, this includes an EST with personnel and equipment, derived from each MDS's maintenance UTCs. *Launch and recover* includes the ability to turn a full combat sortie, which adds the ability to handle all armaments and munitions and address any mission-specific needs, but cannot support long flying days nor repair broken aircraft. This adds to the EST package an ADVON team, also derived from each MDS's maintenance UTCs—again, regardless of the duration. Both of these maintenance capabilities provide flight-line support, and are appropriate for locations where aircraft would not remain overnight.

The *repair* selection for maintenance capability is more appropriate for a beddown location where aircraft will remain overnight for one or more nights. This enables maintenance crews to accomplish more-intensive tasks on the aircraft or its components, not simply turn the aircraft to launch additional sorties. Thus, repair includes one or more backshop maintenance capabilities. Selecting anywhere from one to seven days provides maintenance capability to sustain robust sorties with remove-and-replace repair capabilities, up to but not including deeper diagnostics. Selecting eight to 30 days adds personnel and equipment for deeper diagnostic and repair, such as wheel and tire buildup and NDI. Finally, any value greater than 30 days adds the full scope of capabilities in a typical maintenance UTC (scaled to the number of aircraft), including phase maintenance, AGE fleet maintenance and periodic inspection, and more. These choices enable a user to shape a deployment for a range of situations, rather than simply traditional beddown operations.

Picking up again with the right-hand column in Table 3.4, there are weapon system and flying profile inputs. MDS include Air Force fourth- and fifth-generation fighters, as well as proxies for Marine Corps (F/A-18) and Japanese (F-15J) fighters; tanker and cargo aircraft; and special operations forces (SOF) aircraft.[32] Virtually any number of aircraft can be chosen. Because the personnel and equipment rules are formulas, rather than directly drawing from UTCs, this allows a user to deploy packages much smaller than, or simply different from, standard UTC packages.

Sortie rate is sorties per aircraft per day. Sortie rate and ASD are combined to estimate the flying window. A warning appears if the combination of inputs results in a flying day greater than 14 hours, a general rule of thumb. The sortie rate is used to determine daily fuel consumption, fuels equipment, munitions consumption and requirements, and a portion of aircraft maintenance equipment and personnel.

Mission type has to do with munitions, and the choices are No Munitions, Chaff/Flare, Air-to-Air, and Air-to-Ground. Each choice drives differences in personnel and, particularly, equipment selected in the support package. Choosing the No Munitions option means only a few munitions personnel are sent to support EOD and SF personnel, but not aircraft operations. Selecting Chaff/Flare sends half a dozen personnel, specific to each MDS. The other two selections drive higher numbers of weapons-loading crews, based on factors derived from the maintenance UTCs. Figure 3.5 shows a screenshot of the Operational Scenario input sheet.

Figure 3.5. Screenshot of Lean-START Operational Scenario Inputs

Duration	
Total Duration (days)	10
Max flying days	1
Maintenance capability	Launch & recover

Weapon System & Flying Profile							
Aircraft	Number	Mission	Sortie rate	ASD	Sorties/day	Personnel	Weight (Stons)
B-2			-		0		
B-52			-		0		
F-16CD	12	Air-to-Ground	1.50	1.00	18	111	66
F-15C	0	No Munitions	-	-	0		
F-15E	0	No Munitions	-	-	0		
F-15J	0	No Munitions	-	-	0		
FA-18	0	No Munitions	-	-	0		
F-22	0	No Munitions	-	-	0		
C-130	0		-		0		
EC-130	0		-		0		
MC-130	0		-		0		
HC-130	0		-		0		
AC-130	0		-		0		
KC-130	0		-		0		
KC-10	0		-		0		
KC-135	0		-		0		

Beddown Planning

Beddown planning consists of information about the location and desired levels of capability. Figure 3.6 shows the input categories and options for beddown planning.

[32] Much of the early development of Lean-START was geared toward then-ongoing CODE analysis, which included PACAF beddowns with joint and coalition aircraft. The FA-18 and F-15J both use the same rule sets and deployment requirements as the F15E because that was the closest available proxy (i.e., fourth-generation, two-engine, air-to-ground-capable fighter).

Figure 3.6. Beddown Planning Inputs

Base Input Parameter Options

Base_Type	Conventional Threat	CBRN Threat
High Capability (MOB/COB)	High	High
Med. Capability (COB/IAP)	Low	Medium
Low Capability (COB)		Low
Austere		

POL Supply	Security_Forces	EOD
Austere	Protect base	Threat-driven
Storage/Fuel provided	Protect airfield	Protect aircraft only
Fuel + Fillstands provided	Protect flightline	Minimal
Full service provided	SF Team	None
	None	

POL Support	Firefighting	CBRN Capability
FORCE + fillstands	Protect aircraft and structures	Threat-driven
FORCE + trucks	Protect aircraft only	Minimal
ABFDS	Extract pilot only	None
FARP	C2 only	
	No support	

Advanced ADR
None
Small
Medium
Large
Very large

NOTES: ABFDS = aerial bulk fuel delivery system; COB = collocated operating base; FARP = forward area refueling point; FORCE = fuels operational readiness capability equipment; IAP = international airport; POL = petroleum, oils, and lubricants; MOB = main operating base.

For *base type*, the main distinction is between austere and the other three. An *austere* base drives equipment and personnel primarily in CE and BEAR. The other selections are related to the airfield management itself, driving the need for air traffic control and liaison positions. The base type nomenclature in parentheses is meant to be illustrative. For example, a high-capability MOB/COB could be a U.S.-operated base or Japanese Self-Defense Force base; a medium-capability COB/IAP could be a large airport with robust airfield support (but no military presence) or a former Soviet bloc country's air base; and a low-capability COB could be in the Philippines. The intent is for a planner using the model to select a category that best approximates the known level of support at the beddown location as a baseline starting point.

POL supply is the level of support present at the base, which could vary separately from the choice of base type above. *Austere* simply means no support is available; *storage/fuel provided* means on-base distribution, dispensation, and additives are needed; *fuel + fillstands* means only additives are needed. Full service provided could be for a U.S.-run base, a high-capability partner, or a location with an into-plane contract.

POL support is the level of support the user wishes to provide. FORCE is the Air Force's current expeditionary fuels equipment capability. The first choice (*FORCE + fillstands*) would provide a fixed capability to provide hot or cold pitting;[33] the second choice (*FORCE + trucks*) provides R-11 fuel trucks for dispensation directly into the aircraft. The latter choice is only appropriate for small aircraft, such as fighters, or very small numbers of C-130 aircraft, given the amount of fuel larger aircraft require.

The ABFDS is designed for aerial delivery of fuel to locations where other methods of transportation are impractical.[34] In the Air Force, a FARP is used primarily in quick-turn support of special operations aircraft, providing a highly efficient way of transferring fuel from aircraft to aircraft in a nonstandard or hostile environment.[35] Currently only Air Force Special Operations Command has the capability to perform FARP, but we include it here as an option.

Conventional threat level offers two choices: *high* (i.e., active threat) and *low* (i.e., no active threat). Risk assessment is a complicated and involved process, and Lean-START makes no effort to replicate that process for conventional threats and security forces to counter those threats.

The *Security Forces* menu allows a user to select which part of the base to protect, from the entire base (the standard baseline) down to none at all. This could be a function of host-nation or other security capabilities already available (e.g., if another service protects the base, but the Air Force protects the airfield itself), or simply risk assessment. For a fast-moving and/or short-duration operation, the security situation may be such that only airfield or flight-line security is deemed necessary. Or, in a situation where a small transient alert capability is required to periodically service incoming aircraft, an SF team could be sent but remain on standby until aircraft arrive, move in to secure the flight line while the aircraft is there, then return to its original position.

The *firefighting* inputs also provide the ability to choose levels of capability, either based on known support or risk tolerance. The choices available correspond to the Air Force's own fire protection planning logic. Table 3.3 and its related text explain the capabilities that correspond to these input choices.

CBRN threat is the level of threat to which the base is vulnerable. For *EOD*, the default choice is *threat-driven*, which simply follows standard CE practices for sizing the EOD deployment based on the CBRN threat level. *Protect aircraft only* focuses on the aircraft but not

---

33 *Hot pit refueling* is refueling an aircraft while the engines are still running. This saves time by avoiding shutting down the engines, doing a through-flight inspection, and risking some kind of part failure when turning the aircraft on again. *Cold pit refueling* is fueling with the engines off, but is done with fueling hydrants, direct refueling stations, skid mounts, and fuel service units other than a fuel truck.

34 Air Force Logistics Management Agency, *AEF Fuels Management Pocket Guide*, Maxwell Air Force Base: U.S. Air Force, undated.

35 Air Force Logistics Management Agency, undated.

broader protection of the population), and sends only one EOD team and one leadership team. *Provide minimal support* sends only a leadership team; *none* sends no EOD support.[36]

The logic for *EM capability* is similar: *Threat-driven* follows standard CE practices for sizing the EM deployment based on the threat level; *minimal* sends one management team; and *none* sends no CBRN support.

Finally, Lean-START includes the option to send *ADR capability*. The Air Force CE community has defined Prime BEEF teams to accommodate each capability level, as well as equipment vehicle packages. We adopt these rules wholesale. A user may select *none* for no ADR capability (the default), or one of the current packages: small, medium, large, and very large. For reference, these packages range from 72 personnel and about 500 tons of equipment in the smallest package to 228 personnel and more than 3,000 tons of equipment for the largest. The personnel requirements use the same Prime BEEF and EOD UTCs as a standard deployment. In case of the need for ADR capability, the same CE personnel already deployed would shift from their day-to-day responsibilities and focus on ADR. Because of this, Lean-START assesses the need for Prime BEEF and EOD personnel separately for normal base operations and ADR operations, and takes the maximum of the two. It thus avoids double-counting.

Figure 3.7 shows a screenshot of the input sheet for beddown planning. The outlined choices are on the left side of the figure; outputs are on the right.

Figure 3.7. Screenshot of Lean-START Beddown Planning Inputs

Enter base input parameters.	
Base_Type	Austere
POL Supply	Austere
POL Support	FORCE + trucks
Conventional Threat	Low
Security_Forces	Protect base
CBRN Threat	Low
EOD	Threat-driven
Firefighting	Protect aircraft and structures
CBRN	Threat-driven
Advanced ADR	None
Working MOG	4

ACS Outputs					
ACS Function	HN Provides?	Personnel	Tons		Cost
POL	No	28	20.4	$	190,242
CE Ops - BEAR	No	83	276.7	$	3,695,218
Firefighting	No	54	93.8	$	1,408,239
EOD	No	9	19.6	$	2,610,023
CBRN	No	2	0.0	$	-
Security Forces	No	201	190.9	$	4,873,554
Medical	No	16	0.8	$	103,117
Communications	No	94	104.4	$	2,895,257
Logistics - APO	No	34	335.4	$	996,531
Airfield Ops	No	5	0.1	$	-
HQ - Staff	No	56	14.1	$	161,272
Munitions	No	0	93.8	$	-
ADR	No	0	0.0	$	-
Total		582	1,150		16,933,454

NOTES: APO = aerial port operations; HN = host nation.

The one input menu missing from the discussion above is Working MOG (maximum-on-ground). *Working MOG* refers to the maximum number of cargo aircraft at an airfield that may be simultaneously offloaded. This selection drives the deployment of personnel and equipment (specifically material handling equipment) within an aerial port that will service cargo aircraft. These requirements show up in the "Logistics—APO" ACS functional area.

[36] In the version of Lean-START described in this report, one threat selection drives EM and EOD. In future versions, the EOD threat and CBRN threat will be separate.

Tailoring

Aside from the input menus described above for beddown planning, which provide a kind of capability-driven tailoring, Lean-START allows a user to perform detailed tailoring for both maintenance and ACS. For maintenance equipment, tailoring can be done at the individual equipment item level. For other support, tailoring is done at the UTC level. Four options are provided to tailor out UTCs or maintenance equipment items: operational contract support (OCS), host nation support (HNS), WRM, and risk.

Figure 3.8 shows a screenshot of the screen where detailed outputs and tailoring are seen.

Figure 3.8. Screenshot of Detailed ACS Outputs and Tailoring

ACS UTCs and Functions			Rules Outputs			EQUIPMENT TAILORING						
UTC Code	Function	Quantity Rule	Personnel	Tons	Cost	HN	PREPO	OCS	RISK	FINAL	FINAL WT	FINAL COST
4F9J4	CE Ops - BEAR	1	0	1.1	$ 40,831					1.0	1.1	$ 40,831
4F9J5	CE Ops - BEAR	1	0	1.6	$ 15,920					1.0	1.6	$ 15,920
4FPFN	Firefighting	1	1	0	$ -					1.0	-	$ -
4FPFP	Firefighting	1	6	0	$ -					1.0	-	$ -
4FPWD	CBRN	1	2	0	$ -					1.0	-	$ -
9LRCG	HQ - Staff	1	1	0	$ -					1.0	-	$ -
FFGRL	Medical	1	4	0	$ -					1.0	-	$ -
JFDE2	POL	1	3	0.4	$ 3,156					1.0	0.4	$ 3,156
JFDGE	POL	1	0	0.5	$ 5,114					1.0	0.5	$ 5,114
RFSEM	HQ - Staff	1	0	3.2	$ 32,178					1.0	3.2	$ 32,178
RFSEW	HQ - Staff	1	0	3.9	$ 1,937					1.0	3.9	$ 1,937
RFSR9	HQ - Staff	1	1	0	$ -					1.0	-	$ -
UFBAD	Logistics - APO	1	0	2	$ 17,067					1.0	2.0	$ 17,067
UFBJT	Logistics - APO	1	2	0	$ -					1.0	-	$ -
UFBLJ	Logistics - APO	1	0	0.1	$ 853					1.0	0.1	$ 853
UFBPL	Logistics - APO	1	2	0	$ -					1.0	-	$ -
UFM2J	Logistics - APO	1	0	4.5	$ 142,762					1.0	4.5	$ 142,762
UFM38	Firefighting	1	0	11.8	$ 176,377					1.0	11.8	$ 176,377
UFMJ4	Logistics - APO	1	0	5	$ 42,669					1.0	5.0	$ 42,669
UFMPS	Logistics - APO	1	0	12.5	$ 106,672					1.0	12.5	$ 106,672
UFMX6	Logistics - APO	1	1	0	$ -					1.0	-	$ -
XFA23	CE Ops - BEAR	1	0	2.1	$ 34,000					1.0	2.1	$ 34,000
XFAMU	CE Ops - BEAR	1	0	4.1	$ 112,350					1.0	4.1	$ 112,350

NOTES: PREPO = prepositioning; WT = weight.

On the left side of the figure are the UTCs, followed by the functional area that corresponds to that UTC, then the quantity of the UTC that particular run of the model has demanded (shown here as *quantity rule*), followed by the personnel, tons (i.e., weight of the equipment), and cost (investment cost for those equipment items) for each UTC. The next four columns allow for equipment tailoring. *HN* is for host nation support, *PREPO* is for prepositioned equipment, *OCS* is for operational contract support, and *RISK* provides the user the option to simply do without that capability (i.e., taking the risk of needing the capability but not having it). The last three columns are for final values (quantity, weight, and cost) after incorporating the tailoring inputs.

Results

Results are shown in several sheets. For operations and maintenance, the model shows personnel and weight for aviation, maintenance, and munitions maintenance, by aircraft type. For ACS, it shows personnel and weight by functional area. Finally, Lean-START provides a roll-up of weight and personnel and translates those to equivalents of a transportation mode of the user's choice, e.g., C-17, C-130, trucks, etc. The available modes and vehicles can be changed. Where available, standard planning factors are used.[37] Figure 3.9 shows this summary rollup in three sets of rows: *Ops/Mx* includes aviation, maintenance, and munitions maintenance; *ACS* includes all other support; *Total* provides the sum of these two.

Figure 3.9. Lean-START Aggregate Outputs

Transport Mode: C-17

		Personnel	C-17	Equipment Cost (M) One-Time Investment	Equipment Cost (M) Annual Sustainment
Ops/Mx	Personnel	531	5.9		
	Total Tons	304	6.8	19.7	
	Pallets	76	4.2		
	Rolling Stock Tons	204	4.5		
	TOTAL		14.6		
ACS	Personnel	802	8.9		
	Total Tons	2,480	55.1	42.5	
	Pallets	483	26.8		
	Rolling Stock Tons	707	15.7		
	TOTAL		64.0		
TOTAL	Personnel	1,333	14.8		
	Total Tons	2,783	61.9	62.2	
	Pallets	558	31.0		
	Rolling Stock Tons	910	20.2		
	TOTAL		76.7		

NOTE: Numbers do not add up exactly due to rounding.

For each category, the output shows personnel, total equipment weight in tons, then two subsets of equipment. Pallets are the total pallet positions of palletized equipment; Rolling Stock Tons shows the weight of nonpalletized equipment—i.e., rolling stock, which includes motorized vehicles and nonmotorized but wheeled items, such as generators, that must be towed or pushed around. These outputs are translated to equivalents of the transport mode chosen (yellow input box at the top of Figure 3.9). Palletized versus rolling stock weight and dimensions can matter greatly for such aircraft as the C-17 that has a very high weight capacity relative to its pallet-carrying capacity.

Finally, the model also estimates the one-time investment cost of the equipment and the annual sustainment costs to keep them in storage. To develop investment cost estimates, we married item-level data for each UTC from the LOGDET with purchase price data from the

[37] U.S. Air Force, *Air Mobility Planning Factors*, Air Force Pamphlet 10–1403, December 18, 2003.

federal logistics (FEDLOG) database and then aggregated the costs up to the UTC level. Sustainment costs are estimated using planning factors from McGarvey et al, 2010.[38]

Lean-START is generally straightforward and intuitive. It works for a single operating location. For additional bases or excursions, individual runs must be performed and the results copied and pasted to post-process the data.

---

[38] Ronald G. McGarvey, Robert S. Tripp, Rachel Rue, Thomas Lang, Jerry M. Sollinger, Whitney A. Conner, and Louis Luangkesorn, *Global Combat Support Basing: Robust Prepositioning Strategies for Air Force War Reserve Materiel*, Santa Monica, Calif.: RAND Corporation, MG-902-AF, 2010.

4. Example Calculations

In this chapter, we demonstrate some of the model's capabilities with several illustrative calculations. We highlight some of the key drivers of footprint, discuss trade-offs with distributed basing, demonstrate time-phasing of deployment requirements, and show the use of WRM prepositioning to offset footprint.[39] In each series of figures, we show personnel and/or equipment requirements for a range of ACS categories. We generally divide these categories into two broad groupings. The first category comprises aviation, maintenance, and munitions maintenance. As explained in Chapter 3, these capabilities are directly tied to the type, number, and mission of the aircraft chosen by a user. The second category is what we refer to as "nonmaintenance ACS"—i.e., all the other base support functions that are not directly sensitive to aircraft-related selections but are driven by other factors, such as base population, threat level, or preexisting base infrastructure.

Baseline Case: 24-Ship F-16 Deployment

In this series of calculations, we use a baseline case of a 24-ship deployment of F-16s to an austere location. We vary a few key input parameters to explore the effect on deployment footprint.

Aviation, Maintenance, and Munitions Maintenance

Figure 4.1 shows passenger requirements computed with Lean-START for our baseline case.[40] These are the personnel found in UTCs in the three categories of aircraft support discussed in Chapter 3: aviation, maintenance, and munitions maintenance.

---

[39] Note that all calculations shown exclude munitions and resilience measures (e.g., airfield damage repair).

[40] Aviation values exclude primary pilot for each aircraft.

Figure 4.1. Aviation, Maintenance, and Munitions Maintenance Personnel for 24-Ship of F-16s

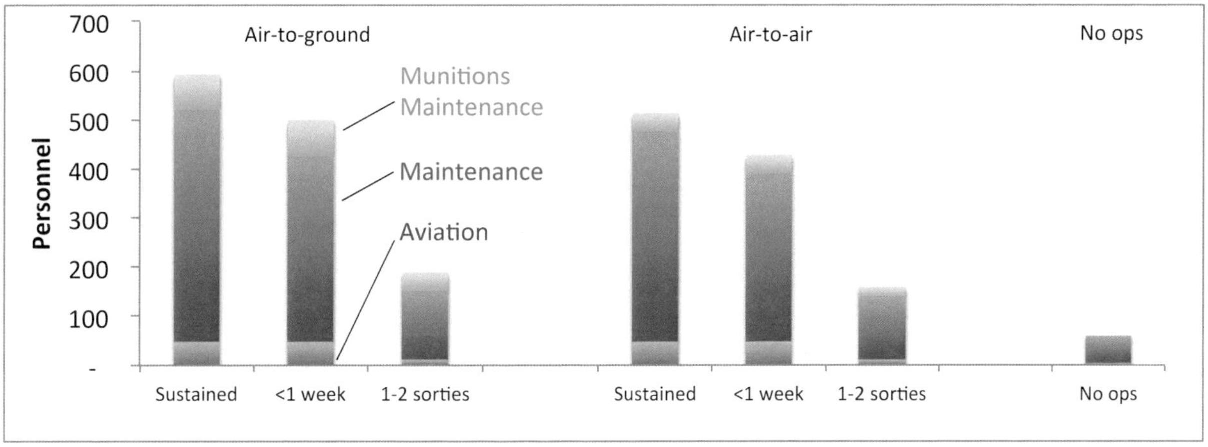

There are three groups of columns in Figure 4.1. On the left side of the figure, the first group uses a mission input of "air-to-ground," and maintenance and munitions maintenance packages include the full range of personnel to support full spectrum operations. The middle group uses a mission input of "air-to-air" munitions only; the last group represents "no operations." The colored column sections, from bottom to top, show operations (aviation) and maintenance personnel (split into aircraft maintenance and munitions maintenance personnel) for each mission type and duration.

Within that leftmost grouping of air-to-ground mission type, duration is varied from "sustained" (the leftmost column), which shows personnel for 30 days or more of operations, to "less than one week" (the second column) to "1–2 sorties" (the final column in the grouping).[41] The second group of three columns shows the same three time frames, but with air-to-air munitions only. The final column on the right side of the figure shows the personnel requirement for no operations, just enough personnel to generate the aircraft to vacate the base.

The biggest reductions in personnel requirements are seen moving from "less than one week" down to 1–2 sorties, and then down to no operations. In the "no operations" case, aviation personnel includes only five operators on the ground (the one pilot per aircraft is not shown because pilots do not count as passengers). For maintenance personnel, the model sends two maintainers per aircraft plus three supervisors, the minimum personnel to generate aircraft to depart.

The reduction in personnel requirements from "less than one week" to "1–2 sorties" is due to a shift away from large-scale, around-the-clock aircraft generation and repair—including long flying windows, multiple shifts, and a 24-hour repair operation that provides for a high daily sortie rate.

[41] The model accommodates a time window of seven to 30 days, but we chose to exclude it for these runs to simplify the figures.

The reduction in personnel requirements from "1–2 sorties" down to "no operations" is a shift from generating at least one full combat sortie (i.e., all weapon systems, munitions, etc.) to simply launching a flyable aircraft. The maintenance footprint drops by more than half because sorties require a larger footprint that includes weapons-loading teams and maintenance to handle munitions buildup.

Additionally, changing the mission type from air-to-ground to air-to-air munitions decreases maintenance personnel requirements by about 10 percent (entirely weapons crews), and halves munitions maintenance crews.

Figure 4.2 shows the same options, but for equipment weight in short tons. We see the same basic pattern across the column heights, but a more pronounced difference between the "less than one week" and "sustained" cases. Operating for one week or less excludes entire capability areas, such as NDI, precision measurement equipment laboratory, and phase inspections, resulting in a smaller equipment footprint.

Figure 4.2. Aviation, Maintenance, and Munitions Maintenance Equipment for 24-Ship of F-16s

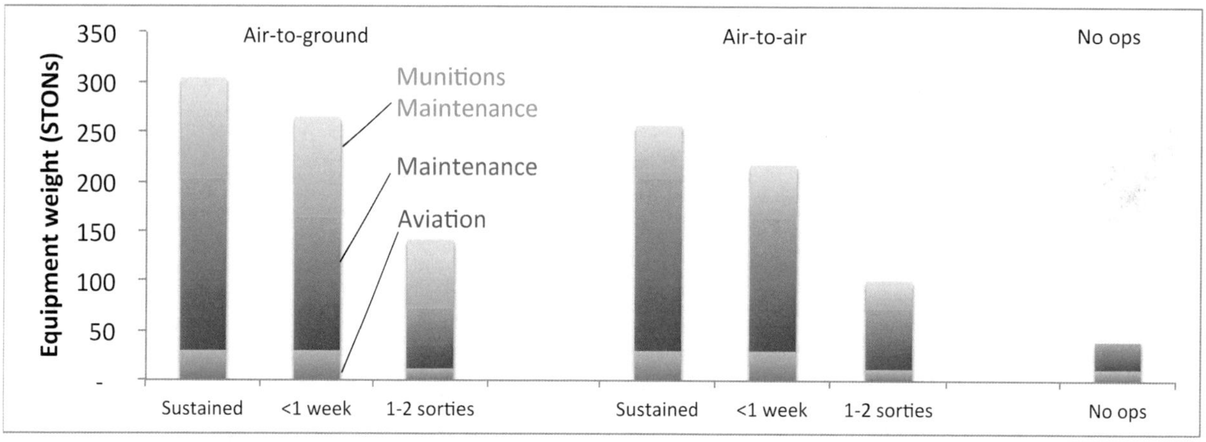

Figures 4.1 and 4.2 show results for a full 24 primary aircraft authorized squadron. We now vary the number of aircraft to see the effect on footprint. Figure 4.3 shows the same options as Figures 4.1 and 4.2, but in this figure, all seven options (i.e., the entire width of the figures) have been compressed into the left third of the figure. The middle and right segments show the same seven options, but for a 12-ship and a six-ship squadron, respectively. Outputs for personnel and equipment are both shown in Figure 4.3, on the top and bottom, respectively, and the three operations and maintenance personnel categories are aggregated together.

Figure 4.3. Footprint Varies With Number of Aircraft

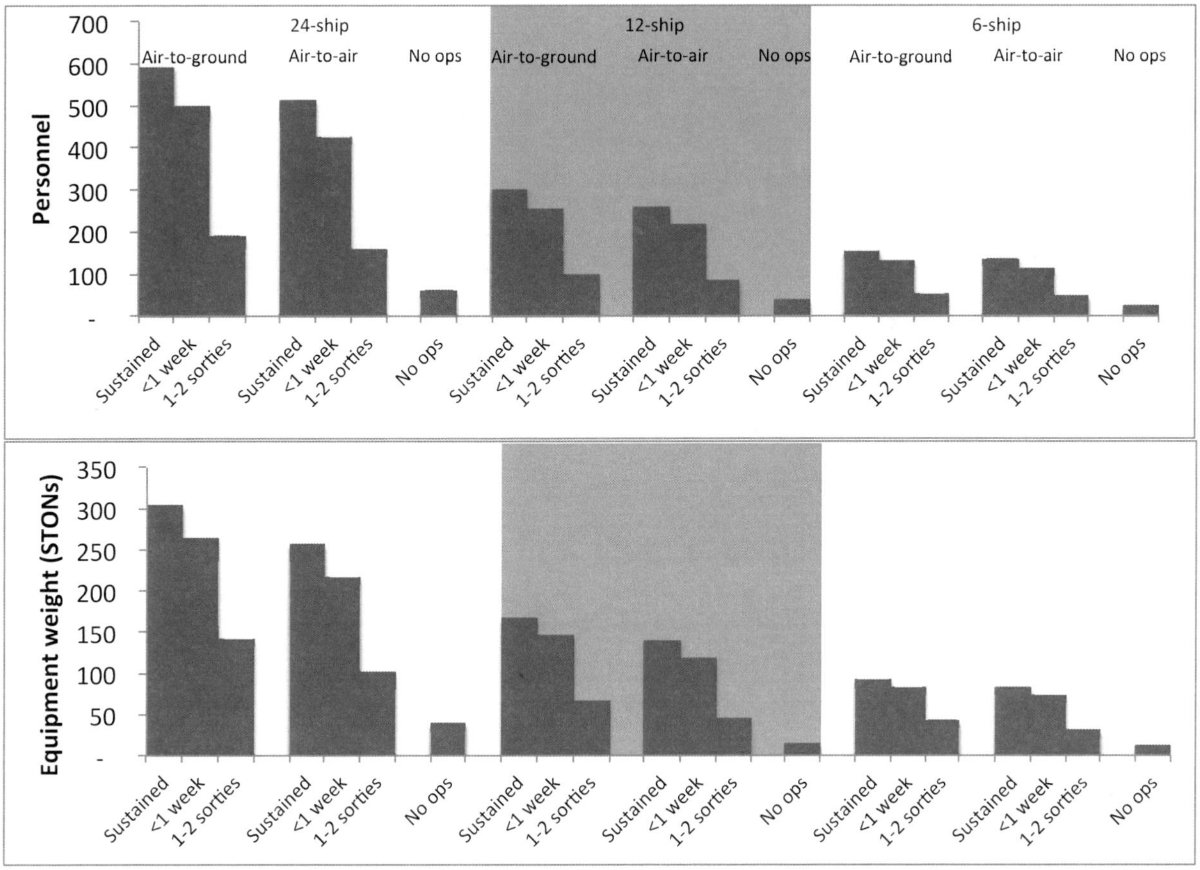

We now start to see a fuller range of possibilities in Figure 4.3. A commander may wish to have a large, robust operation to provide as much flexibility and capability as possible, but that comes at a cost—i.e., a reduction in speed and agility. The two key leverage points in this example are duration and aircraft quantity: A smaller package moves faster, and so does one with less staying power. Displays like this can be generated quickly in Lean-START and can directly inform COA development.

Agile Combat Support

We now show nonmaintenance ACS requirements for the same baseline package of 24 F-16s at an austere location. We display these results differently to accentuate the different drivers of ACS requirements. Figure 4.4 shows the personnel requirements to support our baseline package. The bars show individual functional areas; the pie shows the results in aggregate.

Figure 4.4. ACS Personnel Classified According to Key Drivers

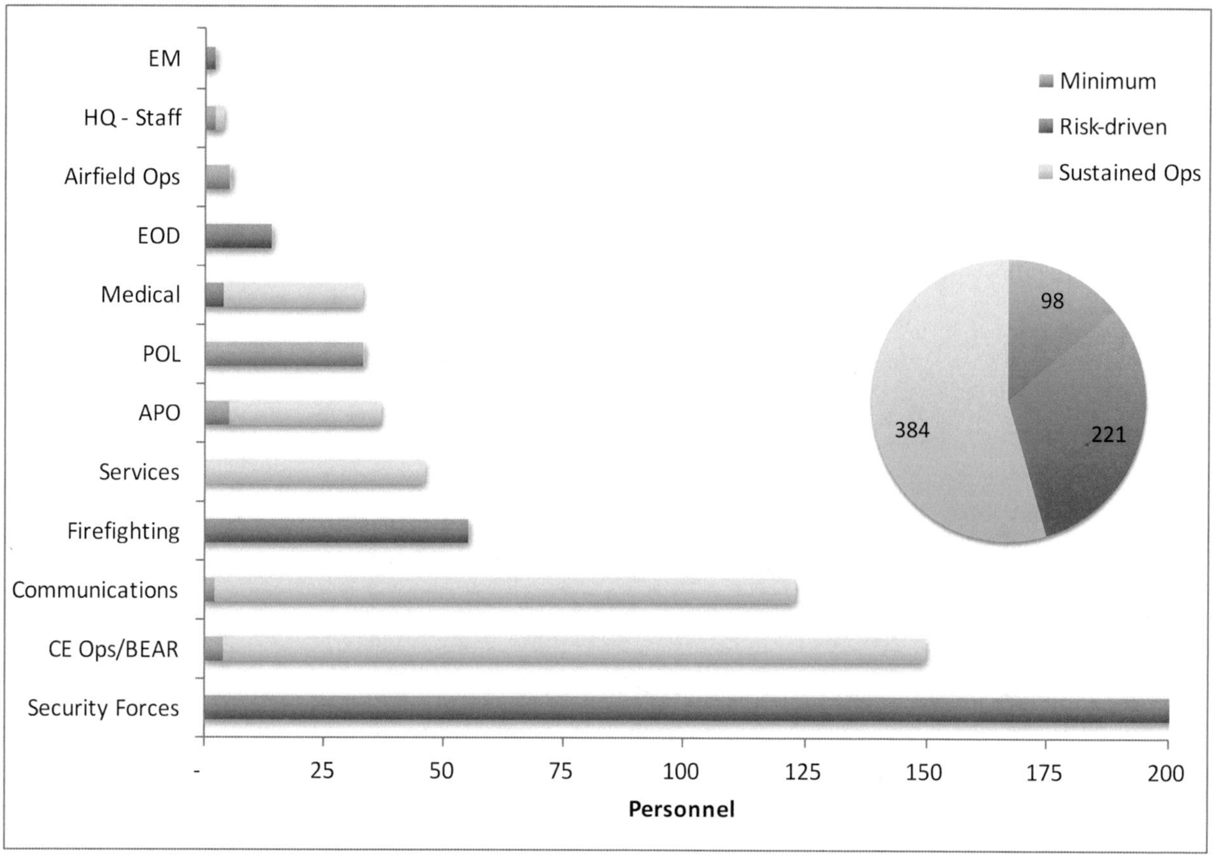

The full length of each bar represents the baseline package at an austere location for a sustained time period—i.e., greater than 30 days. We have shaded sections of these bars to highlight what drives the requirements. The green section of each bar represents the proportion of personnel driven by the need to conduct sustained operations (note that not all bars have green sections); the red represents the proportion of personnel driven by risk; the blue represents the minimum personnel required for sortie generation and base operations.

A personnel requirement driven by sustained operations (green section) predominantly applies to base operating support activities, such as billeting, food, and water. Accordingly, we see larger personnel requirements for CE Ops/BEAR, Communications, Services, APO, and Medical for sustained operations. CE and BEAR include personnel to set up and sustain bare-base equipment, such as shelters (e.g., small tents for personnel and larger ones for maintenance activities), generators, laundry, shower/shave units, and kitchen. Most of the communications personnel are there not to directly support sortie generation, but the larger base population. This requirement virtually disappears if the operation is brief (i.e., one week or less). Services requirements include personnel to prepare and distribute food and water, as well as morale, welfare, and recreation activities.

Medical is a slight exception. The green portion of medical is only partly the direct result of duration. Some medical activities are only needed if the population will be in place for many months (including more routine health needs). When the duration is shortened, those requirements go away. But population is the main driver of expeditionary medical requirements, so, if the duration is shortened, the entire base population drops significantly. Thus, the medical footprint can be reduced from a more robust package to a minimal team.

A personnel requirement driven by risk (red section) means that the key personnel determinant is risk: either the likelihood of a negative outcome or the commander's own risk tolerance. In the model, SF, EOD and firefighting are entirely risk-driven; medical requirements are partly driven by risk. Technically, sorties can be launched with absolutely no firefighting or EOD capability (for example), assuming the commander is comfortable with the trade-off (e.g., risk versus speed or cost). We depict such areas as SF and EOD as entirely risk-driven to emphasize this dimension. But risk tolerance and operational duration are not entirely separable. Commanders may be more willing to accept less capability in areas of safety and security risk mitigation for a short deployment. In the model, we make few risk-based choices based on timing but leave that to the user, providing some of the capability options described in Chapter 4 to make those choices explicit.

Significant risk management is built into many functional areas. If the risk is deemed to be lower, or simply tolerable, less deployed capability is necessary.

The blue section of each bar is the remaining personnel requirement, essentially the minimum footprint necessary to support sortie generation and base operations. This includes a few minimal needs, such as communications to receive the air tasking order and communicate back to the air operations center; limited material handling to offload equipment to initiate operations; air traffic control and airfield management, and, most importantly, fuels. Accordingly, we see blue sections for POL, Airfield Ops, CE Ops/BEAR, APO, Communications, and HQ Staff. Many of these functions may not be required if deploying to a more robust base that has a robust airfield with personnel and equipment, available fuel, etc.

Finally, the pie chart in Figure 4.4 summarizes the share of personnel in each nonmaintenance ACS category. Duration is by far the largest driver of personnel, followed by risk, which shows there is significant leeway in sizing a force package depending on risk tolerance and perception.

Figure 4.5 shows the same analysis as Figure 4.4, but for equipment weight in short tons. The functional areas on the vertical axis are now reordered from heaviest to lightest category; CE Ops/BEAR actually extends far off the horizontal axis to more than 1,200 short tons. A large base population at an austere location with a need for sustained presence drives an enormous footprint.

Figure 4.5. ACS Equipment Weight Classified According to Key Drivers

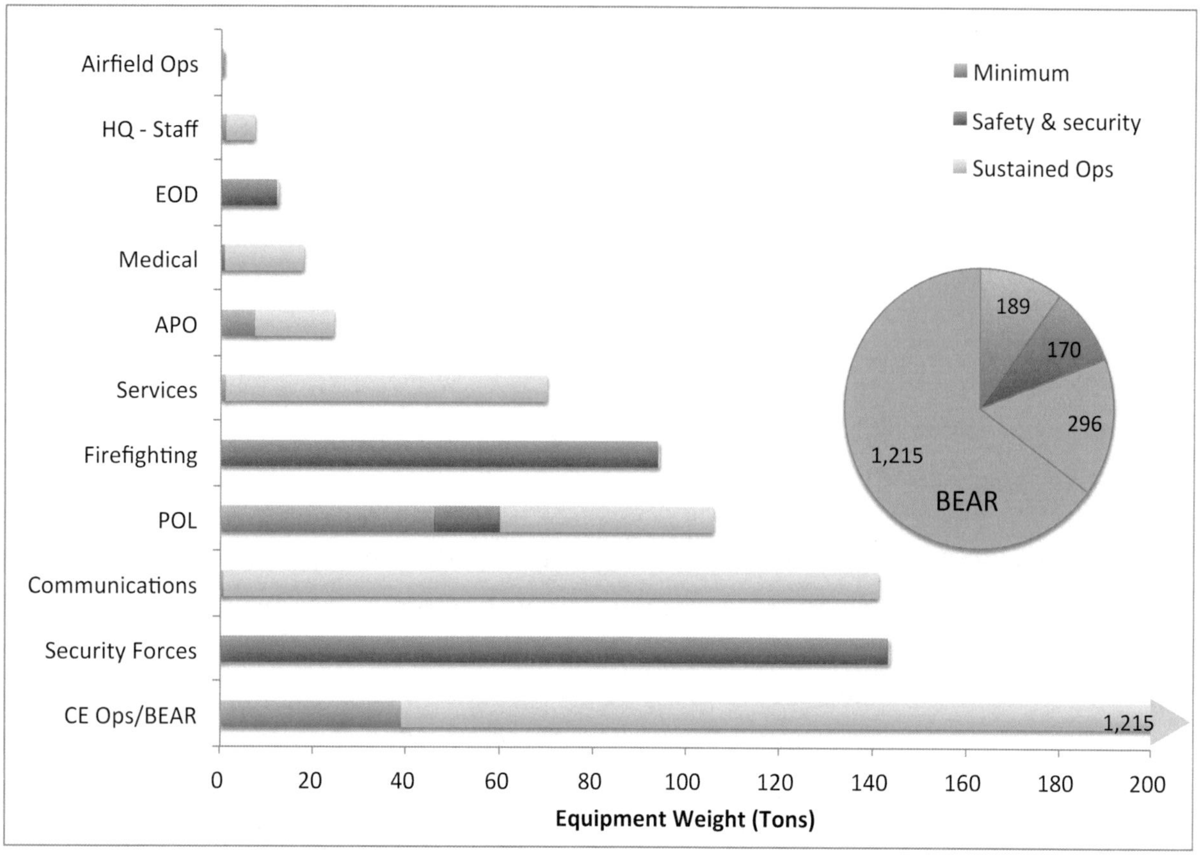

In Figure 4.5, the green section of the pie chart is 80 percent of the total equipment requirements by weight, with three-quarters of that 80 percent being driven by BEAR equipment (shown separately as two green slices of the pie chart). Duration and population drive communications equipment requirements (which reduces with a shorter time frame), as well as firefighting, medical, and many staff function equipment requirements. POL shows a reduction in required equipment from sustained operations to brief operations because we shifted fuel dispensation from fillstands to R-11 fuel trucks. With a shorter duration, the time required to set up fillstands might be deemed prohibitively long.

We see the same risk-driven areas in Figure 4.5 as in Figure 4.4, though with slightly different proportions in some cases. Also, the minimum equipment footprint includes some civil engineering equipment, POL to support sortie generation, and minimal material handling equipment to support equipment offload.

Given the many beddown planning input options discussed in Chapter 4, a wide array of possibilities exists for exploring the footprint implications of duration and risk. One can see just how few personnel and equipment are really required to perform combat operations when the operational concept or conditions dictate such or a commander accepts the risk of reduced capability. This is also a reflection of the standard planning factors assumed in today's Air Force, having evolved to support two decades of sustained operations at predominantly large, robust bases. The tradespace presented by Lean-START suggests that lighter, leaner operations are possible, with proper planning and risk assessment.

Distributed Basing

Much energy is being devoted in the defense community today to increasing the survivability of land-based air forces in contested environments. One feature of some emerging basing concepts is distributing forces at increasing numbers of bases, either to diffuse missile attacks, or to rapidly disperse forces to simply escape damage. It is widely known that more-distributed basing postures increase ACS footprint.[42] Thus, estimating the footprint requirements of distributed basing postures is key to assessing the costs of implementing these strategies.

An important element of any basing strategy is the speed with which the deployment process can be executed. Responsiveness is key to the Air Force's self-described value to the nation's security, so speed-increasing strategies are important to the Air Force's operation.

During the deployment process in the lead-up to a contingency, organically provided equipment (i.e., not from OCS or HNS) is provided from one of three sources: on-site at operating locations, transported from theater prepositioning sites, or transported from CONUS (from unit equipment supplies or CONUS WRM locations). Equipment is prepositioned in-theater largely to offset the need for strategic lift assets (i.e., those transporting from CONUS to overseas operating locations, rather than intratheater lift assets) and to save the time required to wait for those lift assets and make the journey itself. Most Air Force forces are required in the initial days and weeks of a conflict; therefore, much of their equipment is transported by air when possible. Prepositioning Air Force equipment in-theater can thus offset the requirement for strategic airlift assets, primarily C-17s. Thus, when assessing the value or payoff of prepositioned equipment, a common unit of measure is C-17 equivalents.

Unless equipment is stored on-site at its intended point of use (common at some bases in Korea, for example), some transportation assets are still required to move equipment from storage locations to operating locations. This would most commonly be C-130s for airlift, or trucks (e.g., 40-foot tractor trailers) for ground movement. These assets are generally less scarce than C-17s, especially in the early phases of a buildup, so the trade-off is often favorable. Because WRM storage locations are often near theater operating locations, and more-plentiful transportation assets (whether air or ground) can often be more quickly obtained, the delay associated with obtaining equipment from theater storage locations can often be less than deploying that same equipment from CONUS.

---

[42] Patrick Mills, John Drew, John Ausink, Daniel M. Romano, and Rachel Costello, *Balancing Agile Combat Support Manpower to Better Meet the Future Security Environment*, Santa Monica, Calif.: RAND Corporation, RR-337-AF, 2014.

Equipment must obviously be procured, stored, and maintained in working order. Thus, the cost of having prepositioned equipment must be balanced against the benefit garnered in C-17s offset and time saved. Lynch et al. found that in Operation Enduring Freedom and Operation Iraqi Freedom, the majority of ACS equipment deployed to forward operating locations was sent from theater support locations.[43] However, many current basing concepts could levy additional basing and equipment demands—thus, equipment would likely need to be purchased, on top of existing theater supplies.

Estimating the actual time saved by leveraging theater-based WRM is complicated. The delay associated with waiting for strategic airlift, for example, depends on when the equipment must move during the entire deployment and the priorities of queued items. But C-17 equivalents as a metric provide a useful proxy to illustrate the number of planeloads that could be offset by prepositioning equipment in forward support locations.

With that backdrop, Figure 4.6 shows some of the outcomes of distributed basing postures. This figure shows the C-17 equivalents necessary to transport aviation, maintenance, and munitions maintenance equipment to support our baseline 24-ship package.[44]

Figure 4.6. C-17 Equivalents for Aviation, Maintenance, and Munitions Maintenance Equipment Distributed Basing

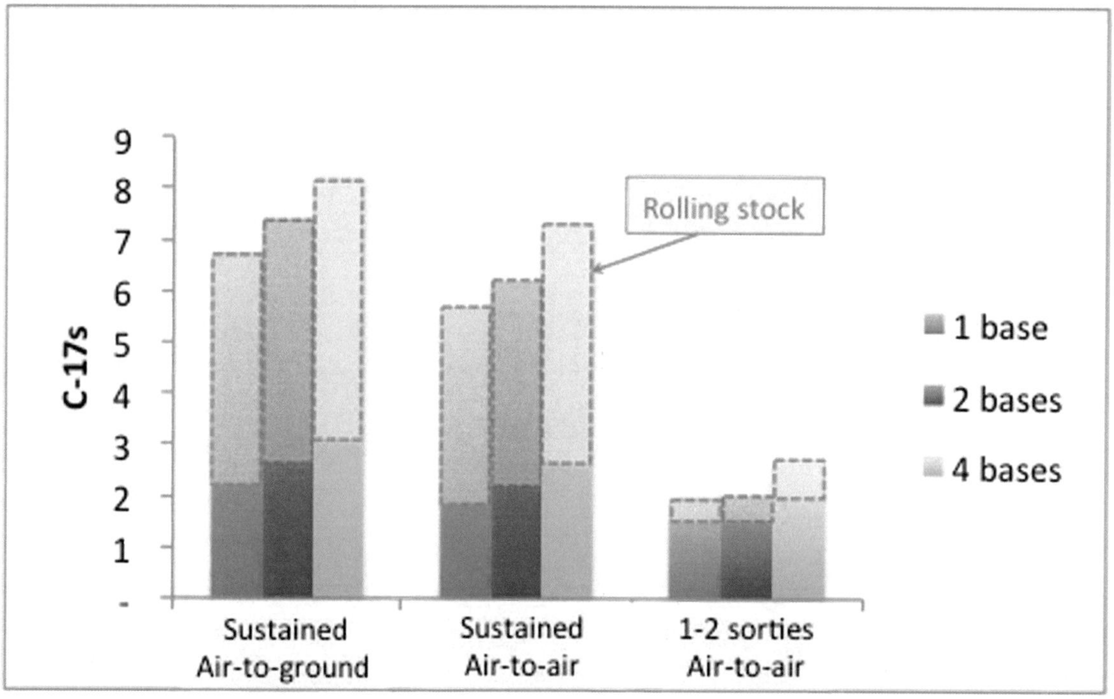

[43] Kristin F. Lynch, John G. Drew, Robert S. Tripp, and Charles Robert Roll, Jr., *Supporting Air and Space Expeditionary Forces: Lessons from Operation Iraqi Freedom*, Santa Monica, Calif.: RAND Corporation, MG-193-AF, 2005.

[44] U.S. Air Force, 2003.

Figure 4.6 shows three sets of three columns: The leftmost grouping represents equipment requirements for sustained air-to-ground operations; the middle represents equipment requirements for sustained air-to-air operations; the right represents equipment requirements for 1–2 sortie generation for air-to-air operations. Additionally, within each set, equipment requirements are shown for the 24-ship package deploying to (from left to right) one base, two bases (a 12-ship at each), and four bases (a 6-ship at each). These results show that dispersing to multiple bases does increase the transportation burden, but fixed costs to support new operating locations are fairly low. Choosing to use only air-to-air munitions offsets some footprint increase, but reducing the sortie generation capacity affects impacts footprint more significantly.

The transparent top sections of each column with dotted borders show the proportion of each package that is rolling stock. We highlight the rolling stock requirement because for maintenance and munitions maintenance, most rolling stock (i.e., vehicles and towed equipment), can be prepositioned in WRM, while most palletized equipment and supplies would not be.[45] Thus, the unit's deployed footprint can be further reduced by incurring the cost of prepositioning WRM, subsequently reducing the cost in terms of strategic airlift assets of a more distributed posture.

Figure 4.7 shows some outcomes of distributed base postures for nonmaintenance ACS equipment requirements (i.e., all the categories shown in Figure 4.5). The three columns within each set show dispersal to a different number of bases—one, two, and four bases. Here, the leftmost set shows the full footprint for an austere location with no prepositioning; the middle set shows the results if maximum prepositioning is used (which comes with a financial cost); the last set shows the results if maximum prepositioning is used and remaining risk-related requirements are eschewed.

---

[45] Much of palletized equipment is not a good candidate for overseas WRM. Spare parts, for example, are not economical to store where an active unit cannot use them. Also, many of the pallets are full of mobility bins with smaller pieces of diagnostic equipment or toolkits that mechanics usually have themselves; they are accountable for these toolkits and can vouch for their reliability. Some of these practices could be relaxed with a policy change, but would likely require additional investment to ensure contractors maintained them in a ready enough state to take out of the box and use right away.

Figure 4.7. C-17 Equivalents of ACS Equipment for Distributed Basing

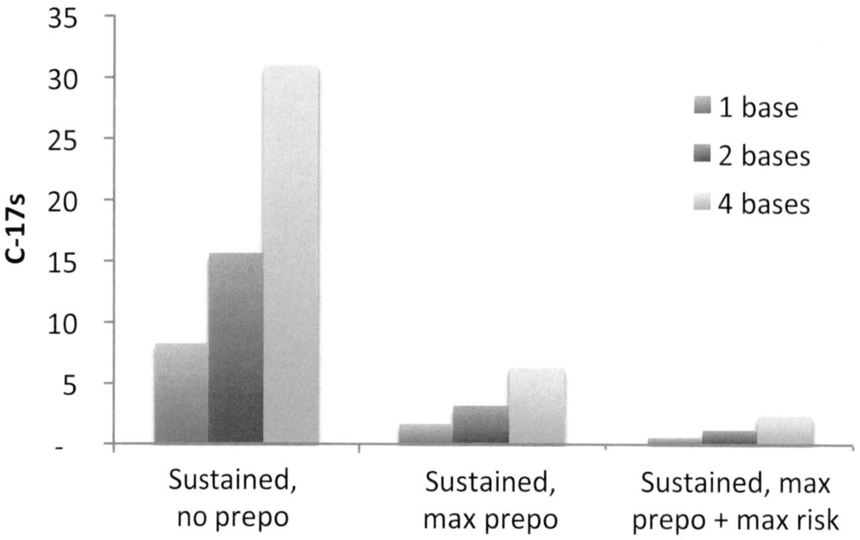

There are three important points to discuss with these model runs. First, the amount of equipment required to open each new base is much larger for nonmaintenance ACS than for aircraft-related requirements. The footprint in each set nearly doubles, then quadruples as the number of bases increases from one to two and from two to four. Distributed basing is extremely costly in terms of ACS equipment.[46] In economic terms, there is a high "fixed cost" for each new operating location. Many ACS functional areas incorporate such a fixed cost, but some areas, including firefighting and security forces, are essentially insensitive to the number of aircraft or base population.

Second, a potentially huge amount (but not all) of that strategic airlift requirement can be offset with prepositioning, depending on where the WRM is located. Third, accepting more risk can further reduce the footprint, requiring only two C-17s instead of six for a four-base posture. At some point, the resolution of the model is challenged, but this gives a good idea of the trade-offs involved.

Time-Phasing of Requirements

Thus far, we have been discussing deployment requirements as one complete movement simultaneously competing for lift. Lean-START can also be used to estimate how transportation requirements could be time-phased to minimize up-front demand on theater or strategic lift assets.

---

[46] This is only a one-week scenario. For sustained operations, much of the support is population-driven, and thus has lower fixed costs and higher variable costs—e.g., BEAR and services. Choosing a sustained duration would slightly dampen the relationship between number of bases and total footprint.

Figure 4.8 shows two sets of calculations for our baseline F-16 package. The left side of the figure shows outputs for aviation, maintenance, and munitions maintenance; the right side, for nonmaintenance ACS. The y-axis is C-17 equivalents. On both sides, blue column sections show demand for personnel; red sections show equipment. The light red dotted sections of each column show the portion of equipment that is rolling stock, showing candidate equipment for prepositioning.

Figure 4.8. Time-Phased Deployment Requirements

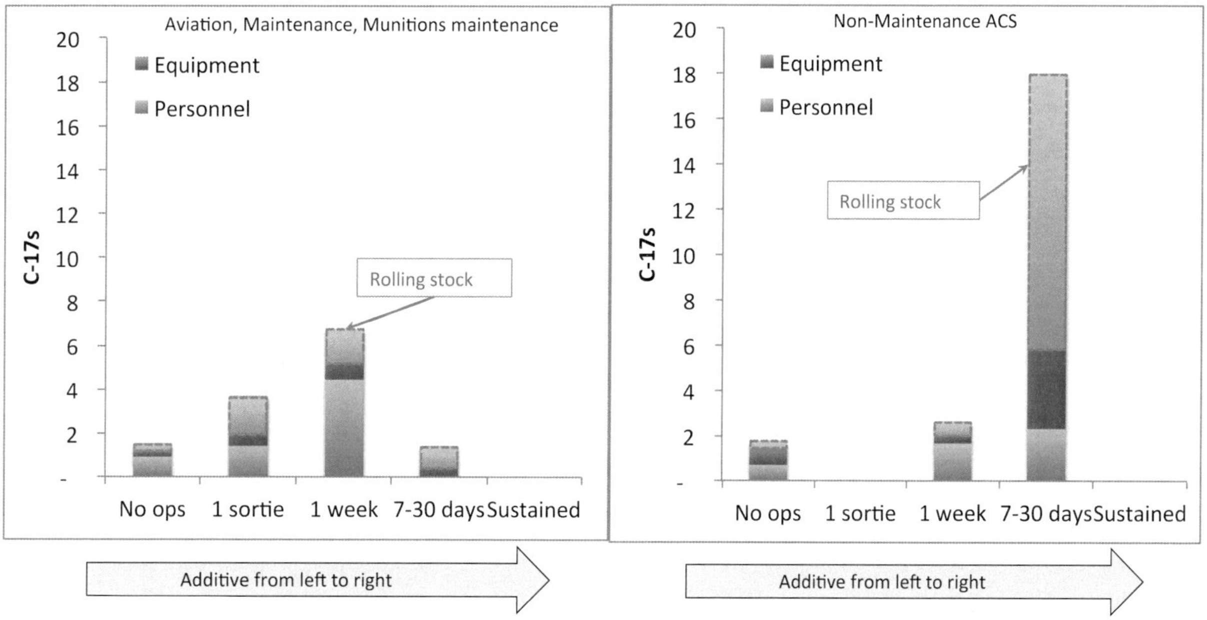

From left to right, the columns are additive. Starting on the left panel of the figure, the leftmost column shows about 1.5 C-17 equivalents are required to meet maintenance requirements for the "no ops" option for our force package (very little of which could be saved with prepositioned equipment).[47] To increase the maintenance capability from evacuation to one or two real combat sorties, another 3.7 C-17 equivalents are required, about 40 percent of which could be offset by prepositioning. Moving from 1–2 sorties to one week of operations requires another 6.7 C-17 equivalents. Very little additional footprint is required to transition to more-sustained operations, which would include deeper maintenance inspections and repairs. Most of that equipment can be prepositioned, but if those resources are not required until day seven or so, lift might then be available, obviating the need to incur such a cost.

The right side of Figure 4.8 has the same outputs, but the pattern is different for nonmaintenance ACS functions. Here, minimal resources (two C-17s) are needed to provide

[47] Very little of the EST package is rolling stock.

initial ACS. Several more are necessary to bring operations from a day to a week, but the real spike is past one week, as base operating support needs expand greatly.

Figure 4.9 looks at the same data slightly differently. Here, the left and right sides of the figure each show aviation, maintenance, and munitions maintenance (blue columns) and ACS (red columns), as well as the total C-17s. The left side of the figure shows the transportation demand in each "phase" if no equipment is prepositioned. The right side shows the values if the maximum possible equipment is placed in WRM.[48]

Figure 4.9. Time-Phased Deployment Requirements

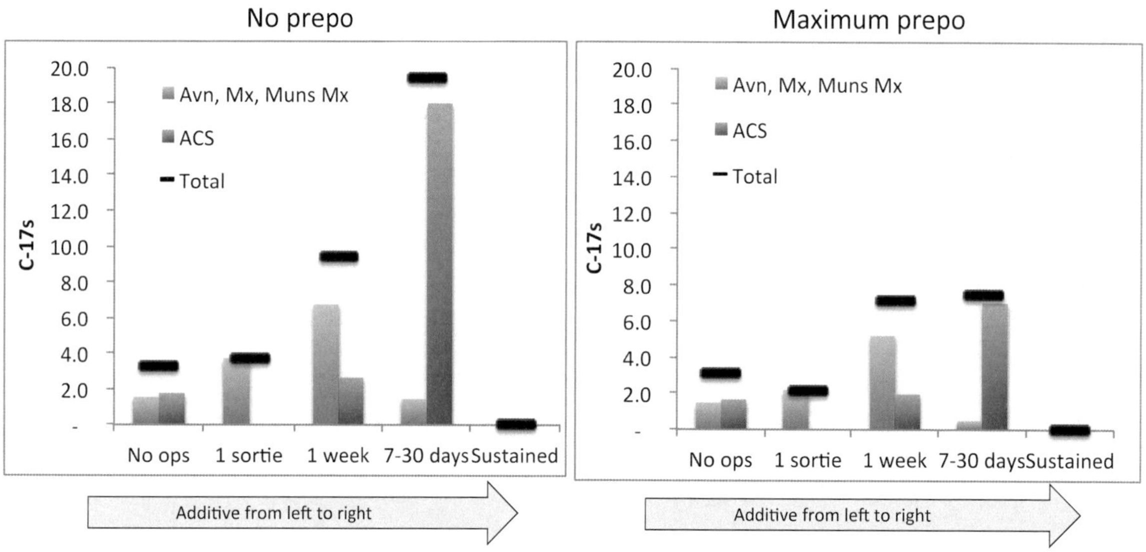

Figure 4.9 shows some of the differences between maintenance and ACS more starkly. Certainly planners and functional SMEs may differ in their opinions about when exactly a certain service ought to be provided, and the commander has the final say on most matters. But one overarching point we extract from this figure is that the "knee in the curve" seems to differ between ops and maintenance relative to ACS, as well as between personnel and equipment, something that can be teased out using the tool.

Prepositioned Equipment Investment Cost

The final example we show more directly targets the trade-offs between deployment footprint and WRM investment cost (as directly output by Lean-START). Figure 4.10 shows calculations for the same baseline example of a 24-ship of F-16s, in this case for one week. The blue line represents full safety and security capability, the red represents no safety and security capability. At the top left of the figure, we see full safety and security capability requires about seven and

[48] The maximum for WRM in this case includes all rolling stock in all functional areas, plus palletized stocks that are expressly intended for WRM, such as BEAR and FORCE.

one-half C-17 equivalents to move the force package's equipment. Moving along the line to the right bottom corner, we see that as more money is invested in WRM prepositioned equipment, the equipment footprint decreases until the entire equipment footprint is stored in WRM. For this example, it takes about $8 million to offset those seven and one-half C-17s. This example is somewhat notional, in that all of the equipment has been included in the calculation—even equipment that, according to current practice, is not placed in WRM. But the basic relationships hold.

Figure 4.10. Trade-Off Between WRM Investment Cost and Strategic Lift Requirement

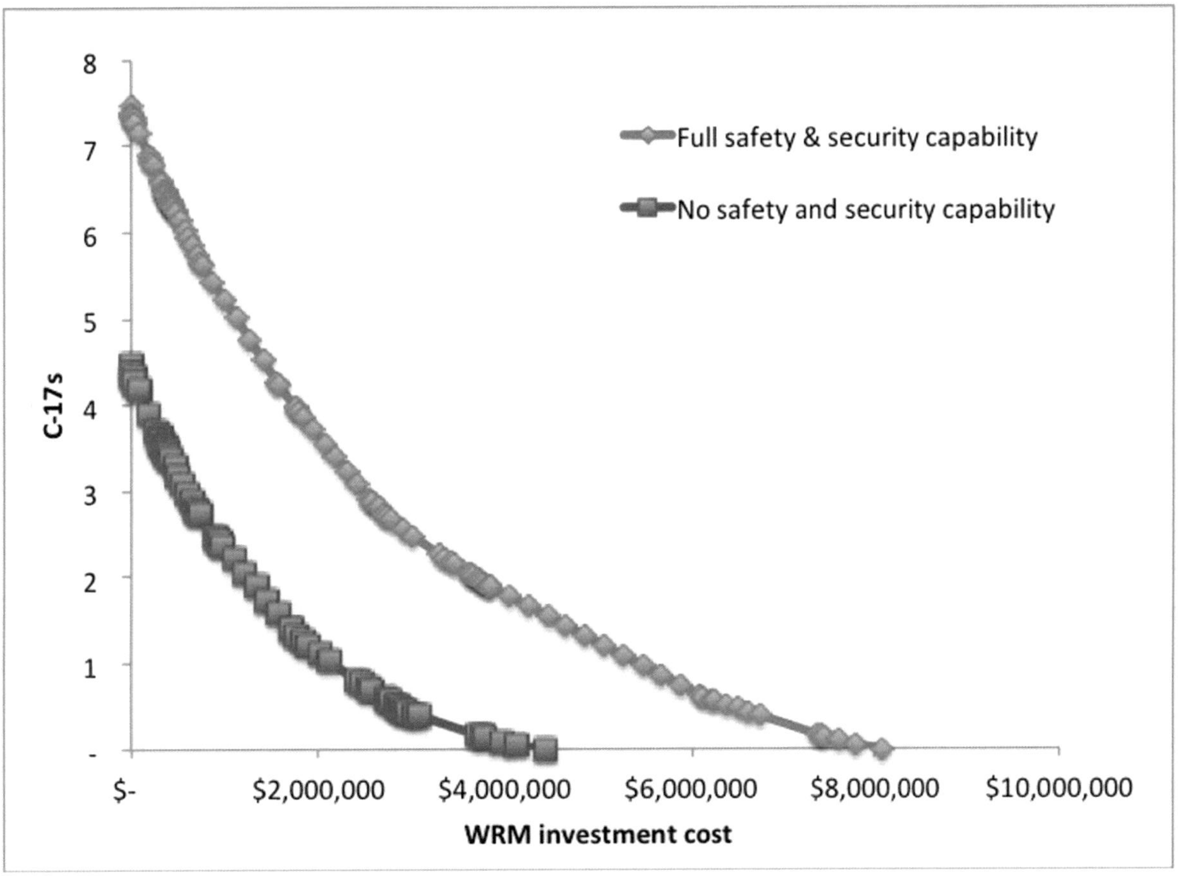

The red line shows the same calculations if all safety and security capabilities are zeroed out; i.e., no SF, EOD, and CBRN. The reduction in C-17 equivalents from full safety and security capability to no capability is dominated by removing SF's heavy vehicles and equipment, followed by EOD heavy equipment package. Creating this display is a bit more involved than most of the previous figures in this chapter, but the same curves can be created for different numbers of aircraft and/or time periods to show the full tradespace.

Risk-Speed Matrix

Table 4.1 returns to the discussion in Chapter 3 regarding risk levels and how to target the best areas to focus on in an effort to increase speed. In this table, we have listed the four risk levels from Joint Publication 5.0 described in Chapter 3. The amount of footprint is derived from our baseline 24-ship package of F-16s operating in an austere environment. If the force package or conditions were to change, the amount of footprint in each area could change, potentially radically. We use this baseline case as a jumping off point to see where footprint can be reduced.

Table 4.1. Risk-Footprint Matrix

	Amount of Footprint		
Risk Level	**Low**	**Medium**	**High**
Extremely High	Communications—Mission APO	Munitions maintenance Fuels	Maintenance
High/Moderate	CBRN[a] EOD[a] Contracting	Communications—Duration Services APO Medical[a] Firefighting[a]	BEAR SF[a] General purpose vehicles
Low	Chaplain Manpower/Personnel Historian		

[a] Safety and security risk.

Across the top are three broad categories expressing the amount of footprint in each functional area—low, medium, and high. We have assigned each functional area (or subset thereof) to a footprint level and risk level. We have noted functional areas identified earlier as having safety and security risks.

Ideally, one would target the areas of most footprint and least risk. Looking to the bottom right, BEAR, SF, and vehicles are the three biggest areas of least risk relative to their overall footprint. Consistent with our earlier discussion, several areas have what we label here as medium footprint, with risk levels of moderate to high, depending on the functional area and conditions. Most of communications and services are driven by the duration of the operation, such that the risk is an issue of quality of life for the personnel. Eventually, lack of services would affect morale and job quality. Medical and firefighting are more issues of risk. If there are no or few incidents, potentially nothing is lost. If there is a base attack or accident, the outcome could be grave. If the duration is short enough, the payoff in saved time might be worth that risk.

5. Conclusions

In this chapter, we summarize the basic ways that Lean-START can support Air Force planners during the planning process. We also highlight some initial findings from our Lean-START analysis of the capability-speed-cost tradespace. Finally, we address some recommendations for future iterations and uses of Lean-START.

Ways to Use Lean-START

The current security environment places a premium on speed and agility—not just with the presentation of forces to counter a threat, but also in the planning and replanning for the operation. Lean-START supports the need for speed and agility of both force employment and planning in a number of ways. First, it allows planners to quickly generate requirements for a scenario. Second, it provides analysis to illuminate the trade-offs among speed, capability, and risks. Finally, it informs COA development and selection.

Using Lean-START, planners are able to quickly determine the combat support capabilities required to support a specified operational scenario. Upon identifying such operational parameters as the number and types of aircraft, the sortie rates for those aircraft, the duration of operations, and the types of missions being flown, Lean-START will use traditional planning factors to generate the quantities of support resources required. This is a capability that existed in Lean-START's predecessor, START.

Beyond the original functions of START, Lean-START also provides planners an opportunity to examine trade-offs. How can speed of deployment be increased by modifying the operational force package or required sortie production capability of the deploying weapon systems? How can speed of deployment be increased by accepting risk to the sortie production capability, or safety and security of the deploying units? What resources that support care and feeding of deployed forces can be eliminated to improve speed? These are questions that get at the fundamental questions of trading capability and risks for speed of deployment.

One of the strengths of Lean-START that enables these analyses is integration of data and information across functional areas. Some of these areas have well-understood tools and deployment rules, others do not. This integration enables a user to grapple with the full range of functional capabilities at once to present a holistic picture of capability, risk, and cost.

The model also integrates information across different types of outputs: weight, personnel, transportation resources, and financial cost. These are all costs of one kind or another, and the model combines them in a way that enables powerful trade-offs to be considered.

Finally, as the Air Force conducts planning in support of a CCDR's campaign, Lean-START enables rapid iteration of COAs that can be provided to the CCDR's planning team. As highlighted in Joint Publication 5.0, planners are often considering and analyzing several courses of action concurrently. This concurrent dynamic planning and replanning is a fundamental element of the adaptive planning and execution framework and is supported by Lean-START's ability to quickly compute deployment requirements and address trade-offs.

Limitations of the Model

We find that two key limitations are informing the complexities of maintenance operations and risk calculus. Tools like these naturally run into the limitation that it is simply difficult to codify in discrete terms many decisions that require either complicated calculations about system interactions or those that require human judgment. One runs into both when planning Air Force deployments. Maintenance is a good example of the complexity issue: Sizing and shaping a maintenance operation with personnel and equipment in the real world requires perhaps a dozen or more inputs, which begins to defeat the express purpose of simplicity in the tool. Security forces are a good example of the risk calculus issues: Assessing risk in a situation, again, depends on many factors, and at this point, only informed human judgment can make a call.

In both cases, however, Lean-START succeeds in bounding the problem. It imagines stark breakpoints where continuous variables really exist, but it can serve to narrow the terms of a debate and explore input parameters in a way that help inform and calibrate human judgment where the decisionmaker lacks deep subject matter expertise in one area or another.

What this translates to is the ability to quickly explore trade-offs like the ones we illustrated in Chapter 4, to hone in on promising or interesting concepts or areas of a problem space, and get to the key drivers—to accept the limitations of reduced capabilities, seek to mitigate the risks, and/or engineer solutions that address those issues more fundamentally.

We believe the tool is best used in the hands of experienced ACS planners, particularly Air Force forces staffs. (This could be logistics planners specifically, or just someone in the job position of planner).

Observations from Initial Lean-START Analysis

The analysis presented in this report is intended to demonstrate the range of calculations that the model can perform. Nonetheless, we do present some observations from our analysis.

Implications for Lean Force Packages

One important question about lean force packages is where the breakpoints occur. For maintenance, the key transition is between 1–2 sorties and more-sustained sortie generation that requires more-robust repair capabilities. Other parameter changes bring increases in requirements (e.g., sortie rate and duration, extended deployments), but the biggest jump seems to be the shift to real repair operations that require a host of additional equipment, personnel for multiple shifts and broader knowledge, and much more robust spare parts kits.

We note that these breakpoints are not exactly the same for larger aircraft, such as C-130s (both mobility and SOF), tankers, and strategic lift aircraft. They simply have different maintenance requirements and concepts, and (maybe most importantly) all have carrying capacity sufficient to haul at least a nominal load of maintenance personnel, equipment, and parts. During discussions with SMEs, it became clear that inside a week or so, most of these larger SOF and mobility aircraft could transport their own operations and maintenance support organically, such that they would only need fuel and nominal airfield support.[49]

Most of the weight and volume of nonmaintenance ACS is driven by base operating support, much of which affects human performance and endurance. Second to base operating support is safety and security, which is often a matter of risk calculus. When operations are focused on combat operations only and/or limited to duration of a week or less, footprint drops drastically. Most of the equipment burdens for short- or long-duration operations are good candidates for WRM prepositioning, but that obviously comes with a cost.

The biggest single potential driver of footprint for safety and security is security forces—both the personnel and equipment, but particularly their vehicles. With shorter operation durations, commanders may be more comfortable accepting more risk, and there are also myriad ways to address the need for security. An obvious one is HNS, which can provide much, if not all, of a base security force's needs. Prepositioning can also reduce an enormous burden.

Also, such emerging concepts as "fight the base" offer ways to mitigate overall demand.[50] Some of the Air Force's footprint is driven by choices in the ways it operates, and it operates in many ways that are different from its sister services. A concept where all airmen are cross-trained and prepared to defend the base (whether holding a rifle or other functions) could go some way to reducing the demand for SF personnel. And finally, such technologies as unmanned aerial vehicles can increase capability and thus reduce some personnel or equipment demands. As stated earlier, the model can reveal some of these drivers and help focus further efforts at research into concepts or technologies.

Implications for Operating in Contested, Degraded, and Operationally Limited Environments

One of the outcomes for emerging basing and operating concepts is what we described above—a sort of envelope within which requirements stay fairly small, and outside of which

---

[49] Discussions with personnel from HQ PACAF on July 26–28, 2016.

[50] Bradley D. Spacy, Edwin H. Oshiba, and Nicholas J. Thomas, "Fight the Base, Recover the Base, Win the War!" *Air and Space Power Journal*, Vol. 30, No. 3, Fall 2016.

requirements tend to balloon. This understanding can shape approaches to evolving support concepts.

Perhaps more significant is the need to view bases as part of an integrated network. The past PAF work we referenced did just that, focusing on the time within which a base could be spun up. The mobility air forces already do that, with the global en-route structure. It is business as usual for cargo and tanker aircraft to transit from location to location across a well-understood global network of bases. Given the potential size of footprint for combat forces, though, incorporating significant maintenance and munitions maintenance requirements even within fairly short durations, much investment of time and perhaps money will be necessary to prepare and manage such a network of bases, even just a regional one.

Current operating concepts are now pushing combat forces to operate in this way. But that cannot be done without more detailed understanding of the base's capabilities, needs, and spin-up time. It is also not possible without excellent coordination of host-nation support, contract support, and equipment prepositioning, all integrated into a coherent strategy. The Air Force is working in this direction, but much remains to be done. A tool like Lean-START can help.

Recommendations and the Way Forward

Lean-START provides a relatively complete capability for determining deployment requirements and assessing various options for trading capability and risks for speed. However, there are additional areas that could be addressed that would benefit the Air Force. We offer the following recommendations:

- **Timing:** One variable that Lean-START does not currently incorporate is timing. The process of preparing, packing, deploying, and initiating operating capability is itself a highly coordinated and orchestrated operation, one with hard constraints and some insurmountable delays. In our discussions with SMEs, timing was a common theme. This needs attention to better flesh out issues associated with packing up and leaving a base, as well as those associated with setting up operations at a base that starts cold—or even warm, with some presence and prepositioned equipment.
- **OCS:** The Air Force has made significant improvement in the way OCS is factored into planning. In its current form, Lean-START can help target OCS operations by highlighting the larger items in a deployment package that could be tailored from the package and provided by OCS. Similarly, future iterations of Lean-START could expedite the planning process by providing the option to identify those items that are known to be available through OCS as part of the scenario creation.
- **Joint basing:** Often, the Air Force is identified as the responsible agent for Base Operating Support Integration (BOSI). In its current form, Lean-START only addresses installation support for Air Force organizations that are being bedded down at a location.

Future iterations could consider the requirements for joint or coalition forces that the Air Force is required to support at the deployed location.

- **Tactics, techniques, procedures (TTPs) and policy:** Lean-START calculations are driven by current TTPs and planning factors. Lean-START also provides an opportunity to examine the effect of those planning factors on the total footprint for force support. The Air Force might consider evaluating current planning factors to determine if they should be adjusted to establish a new baseline for standard levels of support for combat operations.
- **Research and technology:** Lean-START provides the Air Force insight regarding which equipment dramatically increases the deployment footprint and thereby slows the speed of deployment. Those items provide a target of opportunity for research and technology development. The Air Force might consider identifying some of those items and establishing a portfolio of research initiatives designed to provide the same capability at lower costs to the deployment footprint.

Appendix: Air Force Risks Management Policies

Mechanisms for Articulating, Accepting and Managing Risks

The Air Force has different policies and procedures, consistent with joint policy, for risk acceptance and management. Through review of Air Force publications and SME input, we grouped these different policies and procedures into four categories: *waivers, deviations, risk management plans*, and *exemptions*. These categories are defined in the following sections.

Waivers

When units are unable to comply with official policy, guidance, or procedures, units can request a waiver. Waivers are a method for commanders to communicate risk management decisions for a given circumstance or time period where the costs of compliance creates an unacceptable risk to a higher priority, the expected cost of compliance outweighs the benefits, or personnel cannot comply with the requirement due to lack of resources.[51] Approving a waiver implies the commander is accepting the risk created by noncompliance. Waivers are the most formalized process for accepting risk, and are applied to risks felt to be most urgent or critical.

The Air Force has an established system published in Air Force Instruction 33–360, *Publications and Forms Management*, for determining the waiver approval authority and who assumes any risk associated with waivers. Table A.1 lists the criteria for determining the appropriate waiver authority. In Air Force publications and forms, tier numbers may follow stated requirements (e.g., T–0 indicates Tier 0) that provide guidance for determining the appropriate waiver authority when departing from stated resource requirements. For nontiered items with no specified waiver authority, the default approval authority is the publication approval official, through the publication office of primary responsibility (OPR). In our research, we observed that not all Air Force publications and forms have instituted this system of tiered waiver authorities. For example, most of the SF publications reviewed as part of this study specify a waiver authority directly in the text of the document.

Since the last issuance of Air Force Instruction 33–360, the Air Force has renamed some of the organization types that support combat operations. Field Operating Agencies (FOAs) and direct reporting units (DRUs) are now called Primary Support Units (PSUs).

[51] Air Force Instruction 33–360, *Publications and Forms Management,* U.S. Air Force, December 1, 2015.

Table A.1. U.S. Air Force Tiered Waiver Authorities

Tier Number	Publication Level	If Consequence of Noncompliance Is . . .	Or . . .	Then the Waiver Authority Is . . .	And Risk Is Assumed by:
0	Any level	Determined by respective non–Air Force authority (e.g., Congress, White House, Office of the Secretary of Defense, Joint Staff)	N/A	External to Air Force. Requests for waivers are processed through command channels to headquarters Air Force (HAF) publication OPR/HAF functional for submission consideration to non–Air Force authority	N/A
1	Department level only	Noncompliance puts airmen, commanders, or the Air Force strongly at risk of mission or program failure, death, injury, legal jeopardy or fraud, waste, or abuse	Conformity is needed across the Air Force and HAF must be consulted prior to deviations	The requestors' MAJCOM commander, with concurrence of the publication's Approving Official For requests from Air Force Level FOA or DRU, the appropriate FOA or DRU combatant command with concurrence of the publication's Approving Official (processed through the FOA's or DRU's parent HAF two-letter)	MAJCOM commander and concurring HAF publication approving official
2	Departmental, MAJCOM, and Air National Guard levels only	Noncompliance may degrade mission or program effectiveness or efficiency and has potential to create moderate risk of mission or program failure, injury, legal jeopardy or fraud, waste, or abuse	Conformity is needed across the Air Force, but HAF is not needed to be consulted for deviations	The requestors' MAJCOM commander For requests from Air Force Level FOA or DRU, the appropriate FOA or DRU combatant command	MAJCOM commander and requestor
3	Departmental, MAJCOM, FOA, and Numbered Air Force levels	Noncompliance may limit mission or program effectiveness or efficiency and has a relatively remote potential to create risk of mission or program failure, injury, legal jeopardy or fraud, waste, or abuse	Conformity is needed across the Air Force, but commanders may decide to deviate from standard practices	Requesting units' Wing/DRU/FOA/ combatant command (delegable no lower than Group/combatant command or equivalent)	Wing, FOA, or DRU commander

Deviations

Deviations are a temporary adjustment to published guidance that sometimes requires notification to higher authorities.[52] For example, SFs may tailor manpower requirements and physical security measures based on the unique mission needs, terrain, climate, sociopolitical sensitivities, or other factors without requesting a waiver.[53] Guidance is written so that SFs can adopt whatever protection standards are most compatible with the mission requirements and threat. These deviations would normally be identified and discussed (and accepted or rejected) during COA development.

Risk Management Plan

Risk management plans are required when requesting a formal waiver but may also be used without waivers as a separate way to accept and manage risk. The risk management process consists of the following steps: a risk assessment (Risk = Asset Criticality × [Threat × Vulnerability]);[54] a risk tolerance decision;[55] course of action determination; and decision and implementation.

Exemptions

Exemptions are written to bypass guidance and often require some form of approval or verification but may not require a formal waiver.

No Apparent Institutionalized Processes

For some functional areas we did not find explicit institutionalized processes prescribed for articulating, accepting, and managing risk in Air Force publications. These areas are communications, medical, transportation, and services.

Risk Management Policies and Planning

The policies for risk management vary widely by functional area. Table A.2 shows the types of risk policies and procedures exercised for each function area. An "X" indicates we found at least one resource requirement or process within that functional area that uses the policy and procedure category to articulate, accept, and manage risk.

---

[52] Air Force Instruction 10–401, *Air Force Planning and Execution*, U.S. Air Force, December 7, 2006.

[53] Air Force Instruction 31–101, *Integrated Defense,* October 8, 2009, not available to the general public.

[54] Risk management process model from Air Force Instruction 10–245, *Antiterrorism (AT)*, U.S. Air Force, June 25, 2015.

[55] In the case of SF procedures, the Installation Commander is the installation risk taker and must determine his or her risk tolerance.

Table A.2. Functional Areas and Risk Policies and Procedures for Resource Shortfalls

Functional Area	Formal Waiver	Deviation	Risk Management Plan	Exemptions	No Identified Institutionalized Process
Security forces	X	X	X	X	
Fire services	X	X	X		
EOD	X[a]	X	X		
CE Ops/BEAR	X	X			
Munitions	X				
APO	X				
CBRN	X[a]				
Maintenance					X
Munitions maintenance					
Fuels					X
Communications					X
Medical					X
Vehicles					X
Services					X

[a] Emergency Management (Air Force Instruction 10–2501, *Readiness and Emergency Management (R&EM) Flight Operations*, U.S. Air Force, April 20, 2016) does not call out CBRN or EOD directly but states that when compliance is impractical based on local conditions or MAJCOM situations, waivers should be requested through specified Tier Waiver Authority IAW Air Force Instruction 33–360 (2015); for nontier items, waivers requested from the Air Force Civil Engineer through the Air Force Civil Engineering Center.

Generally speaking, the greater the negative impact of the risk, the greater the number of policies focused on managing it. For example, a risk in security forces, fire services, and EOD that becomes reality would have a high negative cost to airmen and resources. Similarly, those areas are also the same areas where some of the greatest airlift savings can be realized by accepting the risk. They might be classified as high-risk, high-reward functional areas in the context of Lean-START.

Discussion

Each functional area considers and accepts or declines risks for different reasons. For example, there are risks that can be approved through waivers that focus on munitions siting, fuel storage, aircraft parking distances, and APO for what loads can be put on an aircraft. The policies governing those decisions may be considered for waiver to improve operational efficiency at the beddown site. For example, locating munitions storage or fuel storage closer to the flight line may increase the speed with which aircraft can be prepped for a mission.

Lean-START focuses on other risk-based decisions that can be made about deployment and beddown that result in bringing less of something—people or equipment—to increase speed of deployment. The commander may decide to eliminate fire trucks or SF vehicles, knowing that each carries a safety or security risk, for the purpose of reducing the number of short tons of the total package being deployed and thereby increasing the speed with which the force can be deployed.

Acknowledgments

Numerous people within the Air Force provided valuable assistance to and support of our work. They are listed here with their rank and position as of the time of our research. We thank Lt Gen Lee Levy, commander of the Air Force Sustainment Center; Maj Gen Theresa Carter, then–commander of the Air Force Installation and Mission Support Center; and Maj Gen Bradley Spacy, then–Director of Expeditionary Support, Air Force Installation and Mission Support Center for sponsoring this work. We would also like to thank Col Marc Vandeveer, Chief, Plans and Analysis Division, Air Force Installation and Mission Support Center/XZP, and Lynn Arias, Air Force Sustainment Center/LGXA, for their support throughout this research. We also thank their colleagues and staffs within the Air Force Installation and Mission Support Center and Air Force Sustainment Center for their time and support.

We also thank the staffs of HQ Pacific Air Forces; HQ U.S. Air Forces, Europe; HQ Air Mobility Command; and HQ Air Combat Command—particularly the logistics, maintenance, and installations personnel, who patiently explained Air Force deployment requirements to us, sometimes more than once.

Many others within and outside the Air Force also helped and supported us; there are too many for us to mention by name. Their ideas and critiques have helped shape this analysis into its current form.

At RAND, we would like to specifically thank Robert S. Tripp and John G. Drew for useful discussions and helpful inputs.

That we received help and insights from those acknowledged here should not imply that they concur with the views expressed in this report. Responsibility for the content of the document, analyses, and conclusions lies solely with the authors.

Abbreviations

ABFDS	air bulk fuel delivery system
ACC	Air Combat Command
ACS	agile combat support
ADR	airfield damage repair
ADVON	advanced echelon
AEF	aerospace expeditionary force
AGE	aerospace ground equipment
AMC	Air Mobility Command
APEX	Adaptive Planning and Execution
APO	aerial port operations
ASD	average sortie duration
BEAR	base expeditionary airfield resources
C2	command and control
CBRN	chemical, biological, radiological, nuclear
CCDR	combatant commander
CE	civil engineering
COA	course of action
COB	collocated operating base
CODE	combat operations in a denied environment
CONUS	continental United States
CDO	contested, degraded, or operationally limited
CSPEC	combat support planning execution and control
DoD	U.S. Department of Defense
DRU	direct reporting unit
EAF	Expeditionary Aerospace Force
EM	emergency management
EOD	explosive ordinance disposal
EST	en-route support team
FARP	forward area refueling point
FOA	field operating agency
FORCE	fuel operational readiness capability equipment
HAF	Headquarters Air Force

HN	host nation
HNS	host nation support
HQ	headquarters
I&MS	installation and mission support
IAP	international airport
IAW	in accordance with
JPEC	joint planning and execution community
Lean-START	Lean-Strategic Tool for the Analysis of Required Transportation
LOGDET	logistics detail
MAJCOM	Major Command
MANFOR	manpower force packaging system
MEFPAK	manpower and equipment force packages
MDS	mission design series
MISCAP	mission capability
MOB	main operating base
NDI	Nondestructive inspection
NFPA	National Fire Protection Association
OCS	operational contract support
OPR	office of primary responsibility
ops	operations
OPT	operational planning team
PACAF	Pacific Air Forces
PAF	Project AIR FORCE
POL	petroleum, oils, lubricants
Prime BEEF	Prime Base Engineer Emergency Force
SECDEF	Secretary of Defense
SF	security forces
SME	subject-matter expert
SOF	special operations forces
START	Strategic Tool for the Analysis of Required Transportation
USAFE	U.S. Air Forces, Europe
UTC	Unit Type Code
WMP-1	War and Mobilization Plan-1
WRM	war reserve materiel

References

Air Force Instruction 10–401, *Air Force Planning and Execution*, U.S. Air Force, December 7, 2006.

Air Force Instruction 10–245, *Antiterrorism (AT)*, U.S. Air Force, June 25, 2015.

Air Force Instruction 10–2501, *Readiness and Emergency Management (R&EM) Flight Operations*, U.S. Air Force, April 20, 2016.

Air Force Instruction 33–360, *Publications and Forms Management,* U.S. Air Force, December 1, 2015.

Air Force Logistics Management Agency, *AEF Fuels Management Pocket Guide*, Maxwell Air Force Base: U.S. Air Force, undated.

Amouzegar, Mahyar, Ronald G. McGarvey, Robert S. Tripp, Louis Luangkesorn, Thomas Lang, and Charles Robert Roll, Jr., *Evaluation of Options for Overseas Combat Support Basing*, Santa Monica, Calif.: RAND Corporation, MG-421-AF, 2006. As of February 24, 2017: http://www.rand.org/pubs/monographs/MG421.html

Amouzegar, Mahyar, Robert S. Tripp, Ronald G. McGarvey, Ed Chan, and Charles Robert Roll, Jr., *Supporting Air and Space Expeditionary Forces: Analysis of Combat Support Basing Options*, Santa Monica, Calif.: RAND Corporation, MG-261-AF, 2004. As of February 24, 2017: http://www.rand.org/pubs/monographs/MG261.html

Brown, Charles Q., Jr., Bradley D. Spacy, and Charles G. Glover III, "Untethered Operations Rapid Mobility and Forward Basing Are Keys to Airpower's Success in the Antiaccess/Area-Denial Environment," *Air & Space Power Journal*, May–June 2015.

Dammeier, David, Meka Toliver, and Logan Smith, "Future Concepts: Overcoming a Power Projection Problem," Air Force Civil Engineering Center, website, undated. As of February 27, 2017: http://www.afcec.af.mil/News/Publications/CE-Online/Future-Concepts-Power-Projection/

Harper, Jon, "U.S. Air Force Preparing for 'High Volume' Operations in Europe," *National Defense Magazine*, April 5, 2016. As of February 27, 2017: http://www.nationaldefensemagazine.org/blog/lists/posts/post.aspx?ID=2146

Joint Publication 1–02, *Department of Defense Dictionary of Military and Associated Terms*, U.S. Department of Defense, November 8, 2010.

Joint Publication 5–0, *Joint Operational Planning,* U.S. Department of Defense, August 11, 2011.

Leftwich, James, Robert S. Tripp, Amanda B. Geller, Patrick Mills, Tom LaTourrette, Charles Robert Roll, Jr., Cauley von Hoffman, and David Johansen, *Supporting Expeditionary Aerospace Forces: An Operational Architecture for Combat Support Execution Planning and Control*, Santa Monica, Calif.: RAND Corporation, Santa Monica, Calif.: RAND Corporation, MR-1536-AF, 2002. As of February 27, 2017: http://www.rand.org/pubs/monograph_reports/MR1536.html

Lynch, Kristin F., John G. Drew, Robert S. Tripp, and Charles Robert Roll, Jr., *Supporting Air and Space Expeditionary Forces: Lessons from Operation Iraqi Freedom*, Santa Monica, Calif.: RAND Corporation, MG-193-AF, 2005. As of February 27, 2017: http://www.rand.org/pubs/monographs/MG193.html

Lynch, Kristin F., John G. Drew, Robert S. Tripp, Daniel M. Romano, Jin Woo Yi, and Amy L. Maletic, *Implementation Actions for Improving Air Force Command and Control Through Enhanced Agile Combat Support Planning, Execution, Monitoring, and Control Processes*, Santa Monica, Calif.: RAND Corporation, RR-259-AF, 2014a. As of February 27, 2017: http://www.rand.org/pubs/research_reports/RR259.html

———, *An Operational Architecture for Improving Air Force Command and Control Through Enhanced Agile Combat Support Planning, Execution, Monitoring, and Control Processes*, Santa Monica, Calif.: RAND Corporation, RR-261-AF, 2014b. As of February 27, 2017: http://www.rand.org/pubs/research_reports/RR261.html

McGarvey, Ronald G., Robert S. Tripp, Rachel Rue, Thomas Lang, Jerry M. Sollinger, Whitney A. Conner, and Louis Luangkesorn, *Global Combat Support Basing: Robust Prepositioning Strategies for Air Force War Reserve Materiel*, Santa Monica, Calif.: RAND Corporation, MG-902-AF, 2010. As of February 27, 2017: http://www.rand.org/pubs/monographs/MG902.html

Mills, Patrick, John Drew, John Ausink, Daniel M. Romano, and Rachel Costello, *Balancing Agile Combat Support Manpower to Better Meet the Future Security Environment*, Santa Monica, Calif.: RAND Corporation, RR-337-AF, 2014. As of February 27, 2017: http://www.rand.org/pubs/research_reports/RR337.html

"Rapid Raptor Package," *Air Force Magazine*, September 27, 2013. As of February 27, 2017: http://www.airforcemag.com/Features/Pages/2013/September%202013/box092613rapid.aspx

Snyder, Don, and Patrick Mills, *Supporting Air and Space Expeditionary Forces: A Methodology for Determining Air Force Deployment Requirements*, Santa Monica, Calif.: RAND Corporation, MG-176-AF, 2004. As of February 24, 2017: http://www.rand.org/pubs/monographs/MG176.html

Snyder, Don, Patrick Mills, Manuel Carrillo, and Adam Resnick, *Supporting Air and Space Expeditionary Forces: Capabilities and Sustainability of Air and Space Expeditionary Forces*, Santa Monica, Calif.: RAND Corporation, MG-303-AF, 2006. As of February 24, 2017:
http://www.rand.org/pubs/monographs/MG303.html

Spacy, Bradley D., Edwin H. Oshiba, and Nicholas J. Thomas, "Fight the Base, Recover the Base, Win the War!" *Air and Space Power Journal*, Vol. 30, No. 3, Fall 2016.

Thomas, Brent, Mahyar Amouzegar, Rachel Costello, Robert A. Guffey, Andrew Karode, Christopher Lynch, Kristin Lynch, Ken Munson, Chad J. R. Ohlandt, Daniel M. Romano, Ricardo Sanchez, Robert S. Tripp, and Joseph Vesely, *Project AIR FORCE Modeling Capabilities for Support of Combat Operations in Denied Environments*, Santa Monica, Calif.: RAND Corporation, RR-427-AF, 2015. As of February 24, 2017:
http://www.rand.org/pubs/research_reports/RR427.html

Tripp, Robert S., Lionel A. Galway, Paul Killingsworth, Eric Peltz, Timothy Ramey, and John G. Drew, *Supporting Expeditionary Aerospace Forces: An Integrated Strategic Agile Combat Support Planning Framework*, Santa Monica, Calif.: RAND Corporation, MR-1056-AF, 1999. As of February 24, 2017:
http://www.rand.org/pubs/monograph_reports/MR1056.html

Tripp, Robert S., Kristin F. Lynch, John G. Drew, and Robert G. DeFeo, *Improving Air Force Command and Control Through Enhanced Agile Combat Support Planning, Execution, Monitoring, and Control Processes*, Santa Monica, Calif.: RAND Corporation, MG-1070-AF, 2012. As of February 27, 2017:
http://www.rand.org/pubs/monographs/MG1070.html

U.S. Air Force, *Air Mobility Planning Factors*, Air Force Pamphlet 10–1403, December 18, 2003.

———, *Civil Engineer Supplement to the War and Mobilization Plan-1 (WMP-1) 2016*, Washington, D.C., 2016.